Sustainable Civil Infrastructures

Sustainable Infrastructure impacts our well-being and day-to-day lives. The infrastructures we are building today will shape our lives tomorrow. The complex and diverse nature of the impacts due to weather extremes on transportation and civil infrastructures can be seen in our roadways, bridges, and buildings. Extreme summer temperatures, droughts, flash floods, and rising numbers of freeze-thaw cycles pose challenges for civil infrastructure and can endanger public safety. We constantly hear how civil infrastructures need constant attention, preservation, and upgrading. Such improvements and developments would obviously benefit from our desired book series that provide sustainable engineering materials and designs. The economic impact is huge and much research has been conducted worldwide. The future holds many opportunities, not only for researchers in a given country, but also for the worldwide field engineers who apply and implement these technologies. We believe that no approach can succeed if it does not unite the efforts of various engineering disciplines from all over the world under one umbrella to offer a beacon of modern solutions to the global infrastructure. Experts from the various engineering disciplines around the globe will participate in this series, including: Geotechnical, Geological, Geoscience, Petroleum, Structural, Transportation, Bridge, Infrastructure, Energy, Architectural, Chemical and Materials, and other related Engineering disciplines.

More information about this series at http://www.springer.com/series/15140

Hany Shehata · Chandrakant S. Desai
Editors

Advances in Numerical Methods in Geotechnical Engineering

Proceedings of the 2nd GeoMEast International Congress and Exhibition on Sustainable Civil Infrastructures, Egypt 2018 – The Official International Congress of the Soil-Structure Interaction Group in Egypt (SSIGE)

Springer

Editors
Hany Shehata
Soil-Structure Interaction Group
in Egypt (SSIGE)
Cairo, Egypt

Chandrakant S. Desai
University of Arizona
Arizona, AZ, USA

ISSN 2366-3405 ISSN 2366-3413 (electronic)
Sustainable Civil Infrastructures
ISBN 978-3-030-01925-9 ISBN 978-3-030-01926-6 (eBook)
https://doi.org/10.1007/978-3-030-01926-6

Library of Congress Control Number: 2018957623

This Springer imprint is published by the registered company Springer Nature Switzerland AG
The registered company address is: Gewerbestrasse 11, 6330 Cham, Switzerland

Contents

About the Editors

Founding President, General Secretary and Treasurer of the International Association for Computer Methods and Advances in Geomechanics (IACMAG)

Chandrakant S. Desai is a Regents' Professor (Emeritus), Department of Civil Engineering and Engineering Mechanics, University of Arizona, Tucson, Arizona. He has made original and significant contributions in basic and applied research in material-constitutive modeling, laboratory testing, and computational methods for a wide range of problems in civil engineering related to geomechanics/geotechnical engineering, structural mechanics/structural engineering, mechanical engineering, and electronic packaging. His research on the development of the new and innovative disturbed state concept (DSC) for constitutive modeling of geomaterials and interfaces/joints has found significant engineering applications. In conjunction with nonlinear finite element methods, it provides a new and alternative procedure for analysis, design, and reliability for challenging and complex problems of modern technology. He has authored/edited about 20 books and 19 chapters and has been author/co-author of over 320 technical papers in refereed journals and conferences.

His research contributions have received outstanding recognitions at national and international levels, some of which are identified as: (a) development and applications of finite element method for problems involving interaction between structures and foundations, (b) the thin-layer interface element for simulation of contacts (interfaces and joints), (c) the residual flow procedure for free surface seepage, (d) a novel fundamental approach for microstructural instability including liquefaction, and (e) the disturbed state concept for modeling of engineering materials and interfaces, including thermomechanical and rate-dependent behavior of materials in electronic chip-substrate systems.

His book on the finite element method (Desai and Abel) published in 1972 was the first formal text on the subject in the USA, second in the world. In 1979, he authored the pioneering and the first text for teaching the finite element method to undergraduate students. His book on Constitutive Laws for Engineering Materials (Desai and Siriwardane) in 1984 is considered to be the first on the subject that presented a combination of various material models based on continuum mechanics. In 2001, he authored the book on the Disturbed State Concept (DSC) that presents an innovative concept for modeling materials and contacts in a unified manner, combining the continuum mechanics models and a novel idea for introducing the discontinuities in the deforming material. In 1977, he co-edited (Desai and Christian) including his own contributed chapters, the first book on Numerical Methods in Geotechnical Engineering that deals with problems from geotechnical and structural engineering. In 20013–14, he has co-authored (Desai and Zaman) the book, Advanced Geotechnical Engineering: Soil-structure Interaction using Computer and Material Models, which is unique because of its scope, contents, and connection between research and applications.

He was the founding General Editor of the International Journal for Numerical and Analytical Methods in Geomechanics from 1977 to 2000. He is the founding Editor-in-Chief of the International Journal of Geomechanics, published by Geo-Institute, ASC, 2001–2008, and now he serves as its Advisory Editor. He has served as a member of editorial boards of 15 journals and has been chair/member of a number of committees of various national and international societies. He is Founding President of the International

Association for Computer Methods and Advances in Geomechanics (IACMAG). He is credited with introducing the interdisciplinary definition of Geomechanics that involves various areas such as geotechnical engineering and rock mechanics, statics and dynamics of interacting structures and foundations, fluid flow through porous media, geoenvironmental engineering, natural hazards and earthquakes, landslides and subsidence, petroleum engineering, offshore and marine technology, geological engineering and modeling, geothermal energy, ice mechanics, and lunar and planetary geomechanics.

He has received a number of awards and recognition, e.g., Fellow, National Academy of Engineering, India; Lifetime Achievement Award, Alumni Association of VJTI, University of Bombay; The Distinguished Member Award by the American Society of Civil Engineers (ASCE); The Nathan M. Newmark Medal, by Structural Engineering and Engineering Mechanics Institute, ASCE; The Karl Terzaghi Award, by Geo Institute, (ASCE); Honorary Professor, University of Nottingham, UK; Diamond Jubilee Honor, Indian Geotechnical Society; Suklje Award/Lecture, Slovenian Geotechnical Society; HIND Rattan (Jewel of India) Award, by Non-resident Society, New Delhi, India; Meritorious Civilian Service Award by the US Corps of Engineers; Alexander von Humboldt Stiftung Prize by the German Government; Outstanding Contributions Medal by the International Association for Computer Methods and Advances in Geomechanics; Outstanding Contributions Medal in Mechanics by the Czech Academy of Sciences; Clock Award for outstanding Contributions for Thermomechanical Analysis in Electronic Packaging by the Electrical and Electronic Packaging Division, ASME; Five Star Faculty Teaching Finalist Award and the El Paso Natural Gas Foundation Faculty Achievement Award, at the University of Arizona, Tucson, Arizona.

Hany Farouk Shehata He is the founder and CEO of the Soil-Structure Interaction Group in Egypt "SSIGE." He is a partner and vice-president of EHE-Consulting Group in the Middle East, and managing editor of the "Innovative Infrastructure Solutions" journal, published by Springer. He worked in the field of civil engineering early, while studying, with Bechtel Egypt Contracting & PM Company, LLC. His professional experience includes working in culverts, small tunnels, pipe installation, earth reinforcement, soil stabilization, and small bridges. He also has been involved in teaching, research, and consulting. His areas of specialization include static and dynamic soil–structure interactions involving buildings, roads, water structures, retaining walls, earth reinforcement, and bridges, as well as, different disciplines of project management and contract administration. He is the author of an Arabic practical book titled "Practical Solutions for Different Geotechnical Works: The Practical Engineers' Guidelines." He is currently working on a new book titled "Soil-Foundation-Superstructure Interaction: Structural Integration." He is the contributor of more than 50 publications in national and international conferences and journals. He served as a co-chair of the GeoChina 2016 International Conference in Shandong, China. He serves also as a co-chair and secretary general of the GeoMEast 2017 International Conference in Sharm El-Sheikh, Egypt, 2016 Outstanding reviewer of the ASCE as selected by the Editorial Board of International Journal of Geomechanics.

Constitutive Modeling of Geologic Materials and Interfaces: Significant for Geomechanics

Chandrakant S. Desai(✉)

Department of Civil Engineering and Engineering Mechanics, University of Arizona Tucson, Tucson, AZ 85721, USA

Abstract. Characterization of mechanical behavior of materials and interfaces and joints (contacts), called constitutive modeling, is vital for realistic and economic solutions of geotechnical problems by using conventional and advanced computer methods. Since the latter allows complex factors that influence the material behavior, development and use of general and unified constitutive models that allow for such factors assume enhanced importance. This paper presents a brief review of available constitutive models and their limitations, and introduces and provides details of a unified and powerful modeling approach called the *disturbed state concept* (DSC) that can be used for a wide range of geologic, concrete, asphaltic, metallic, solders and polymeric materials, and contacts. The DSC with the *hierarchical single surface* (HISS) plasticity model has been used successfully for a wide range of geologic materials and contacts, and for prediction of measured behavior for many boundary value problems. Typical examples for validation of the behavior geologic martials and contacts, and of a boundary value problem are presented in the paper.

1 Intruduction

Constitutive modeling for characterization of behavior of geologic materials and interfaces and joints, often termed as contacts, play a vital role for accurate and economic analysis and design of problems in geotechnical engineering using conventional or modern computer methods. The latter can account for many significant factors such as arbitrary and multidimensional geometries, nonhomogeneities in material composition, complex boundary conditions, nonlinear material behavior and complex loading conditions. Constitutive modeling has assumed special importance when used in conjunction with the computer methods which allow solution of challenging and complex problems that were usually not solvable by conventional and closed form procedures.

Constitutive modeling defines the behavior of solids and interfaces under mechanical and environmental loadings and plays perhaps the most important role for realistic solutions. A great number of constitutive models, from simple to the advanced, have been proposed. Most of them account for specific characteristics of the material. However, a deforming material may experience *simultaneously*, many characteristics such as elastic, plastic and creep strains, different loading (stress) paths, volume change under shear stress, microcracking leading to fracture and failure, strain softening or degradation, and healing or strengthening. Hence, there is a need for developing unified models that account for these characteristics, very often simultaneously. The

H. Shehata and C. S. Desai (Eds.): GeoMEast 2018, SUCI, pp. 1–31, 2019.
https://doi.org/10.1007/978-3-030-01926-6_1

constitutive modeling based on the unified disturbed state concept (DSC) which can allow for the foregoing factors simultaneously, is a main objective of this paper.

A brief description of the previous models is given below under Brief History. Then details are given of disturbed state concept (DSC) that provides a unified and hierarchical approach to develop constitutive models for a wide range of solid materials and contacts. A unified plasticity approach called hierarchical single surface (HISS) model, which contains most of the available plasticity models as special cases and used commonly to define the relative intact (RI) component in the DSC, described later, is also described. The resulting model is referred to as DSC/HISS model. For detailed information for previous models, computer procedures, the DSC and applications, number publications cited in [1–4] can be consulted.

1.1 Objectives and Scope

The main objection is to present various aspects of constitutive models such as basic mathematical modeling, parameter identification and determination based on appropriate testing, validation at specimen level and validation at practical boundary value problem level.

Various factors influence the material behavior. Significant among those factors should be included in mathematical definition of constitutive models. Some such factors are (1) Initial or *in situ* stress or strain (2) elastic, irreversible (plastic) and creep deformations, (3) stress path, (4) volume change under shear and its initiation during loading, (5) isotropic and anisotropic hardening, (6) stress (load) path dependence, (7) discontinuities: inherent and induced, (8) microstructural modifications leading to fracture and softening, and microstructural instabilities like failure and liquefaction, (9) degradation or softening, (10) strengthening or healing, (11) forces: loads (static, repetitive and cyclic (dynamic), temperature, moisture (fluid) and chemical effects.

Considerable research activities and number of publications have taken place over the last few decades on a variety of constitutive models. However, many of such models have addressed only specific factors. The DSC is considered to be a unique and unified approach that account for the above factors *simultaneously* and can be applied to a wide range of materials and contacts.

It is interesting to note that behavior of geologic materials like soils, rocks and concrete is affected by many of the above factors compared to other materials like metals and alloys. Hence, developments for advanced constitutive models for such materials have taken place perhaps more actively. Because of its generality and hierarchical nature, the DSC initially developed for geologic materials, has been used successfully for other materials such as metals, alloys, materials in electronic packaging, ceramics, silicon and polymers. Publications on the DSC by the author and coworkers, and many others have taken place in *diverse* publication media (e.g. journals in engineering and physics) for materials such as geologic (soils and rocks), concrete, asphalt concrete, metals, alloys (e.g. leaded and unleaded solders), silicon and polymers, and interfaces and joints. Since reviews of available models have been presented in [1–5], their details are not included herein.

2 Constitutive Modeling- Brief History

A list of constitutive models often used in Geomechanics is given below:

(a) Linear and nonlinear elasticity, (b) Conventional plasticity,

(c) Continuous yield plasticity, (d) Generalized plasticity, (e) Hierarchical Single Surface (HISS) plasticity, (f) Continuous and discontinuous materials:

- Fracture mechanics, Micromechanics,, damage mechanics,

(g) Unified models: Disturbed State Concept (DSC).

A brief descritions of the limitations of various models and need for unified models are given below:

Linear and non nlinear elstic models are valid only for elastic or recoverable strains. They do not allow for irreversible (plastic) strains, and it is difficult to model some loadings, rate of loading, realistic volume change and microcacraking which may cause softening or healing.

Conventional plasticity can inlcude Von Mises, Tresca, Mohr-Coulomb and Drucker Prager models, in which the behavior is elastic until the yield or critical stress when plascitity can ensue. Usually they are valid for ultimate or failure load and do not allow for relaistic volume change and continous yielding, i.e. from the beginning of loading.

Continous yield plasticity, e.g. critical state and cap models [1–6] can account for yileding from the start of the loading, stress path and volume change. However, they do not allow for volume change before the peak stress and diffreent strength (yielding) under diffrenet stress paths.

Generalized plasticity model that can account for nonassociative and anisotropic hardening behavior and the bounding surface models and their modifications have been presented in [7–11]. Sometimes such models based on the continuum assumption have been modified, by using *ad hoc* schemes or external enhancements [12], to handle discontinuous, softening or degradation. They allow for certain factors such as kinematic and anisotropic hardening, but they lack unified and hierarchical framework. Also, a main reason for softening or degradation is considered to be related to discontinuities due to microcracking; hence, such models may have limited validity for materials involving discontinuities. Also, usually the yielding is dependent only on the volumetric response (strains), and they do not allow for volume change before the peak stress.

The Hierarchical Single Surface (HISS) plasticity models account for continous yielding depedent on both volumetric and deviatoric strains, volume change before the peak stress, diffrenet strenghts under diffrenet stress paths, associative, nonassociative, isotropic and anisotropic yileding. The HISS models involve *single continous* yield surfaces, and inlcue most of the previous continuum plasticity models as special cases [1]. However, as all the above models based on the assumption of continuous materail, the HISS models do not allow for the effect of discontinuities.

Most materials and contacts involve discontinuties before and induced during loaidng. The models from elasticity, plasticity, visco-plasticity are based on the assumption of *continous* material. Hence they are not able to account for the dincontinuities, which are improtant for realistic behavior of many materials. Models like fracturemechanics, micromechnaics and damage mechanics allow fot the discontinuities; however, they suffer from some limitations. The frature mechnanics models often assume linear elastic response, and requires *a priori* assumption for location of cracks, which may not be relaistic because microcraks/cracks can occur anywhere within the material depending on factors such as geometry, material compositiion, and constitutive behavior. Micromechnics is based on microlevel mechanisms and very often derives materials parameters from the behavior of "finite" (macrolevel) sized specimens, which may not be relaistic. Also, the process of integration from micro (part) to macro (whole) may not allow properly for the interaction of particles in complex material systems [2]. The models based on classical damage mechanics assume the "damaged" material has no strength, i.e. they ignore the coupling beteween the undamaged and damaged parts, which may lead *to local* models. As the coupling is vital for realistic behavior, some modified damage models introduce external enrichments (12), which may add to complexity.

The foregoing models account for only specific behavioral features like as a patchwork, and cannot handle such features oultined above, simultaneously. Hence, it is necesaary to develop unified constitituve models that can allow various fetaures at a time and can be used for a wide range of materials and contacts. The DSC with HISS plasticity (DSC/HISS) is considered to be perahaps the only unified and general model which has been used successfully for modeling a wide range of materials and contacts. The main objective of this paper to present the DSC with breif descirption its theory, identification and determination parameters from approriate testing, and validation at the speimen and boundary value problem levels. The HISS plasticity which is often used for modeling the relative intact (RI) behavior in the DSC (descibed later) is first presentecd.

2.1 Hierarchical Single Surface (HISS) Plasticity

The need for a unified and general plasticity model led to the development of the hierarchical single surface (HISS) plasticity model [1, 13–15].The unified and hierarchical HISS plasticity model involves a *single and continuous* yield surface, can account for most of the factors listed above. However, it cannot account for discontinuities (inherent and induced), microstructural modifications leading to fracture and softening, and instabilities like failure and liquefaction.

The yield surface, F, in HISS associative plasticity is expressed as (Fig. 1):

$$F = \overline{J_{2D}} - (-\alpha \overline{J_1}^n + \gamma \overline{J_1}^2)(1 - \beta S_r)^{-0.5} = 0 \quad (1)$$

where $\overline{J_{2D}} = J_{2D}/p_a^2$ is the non-dimensional second invariant of the deviatoric stress tensor, $\overline{J_1} = (J_1 + 3R)/p_a$, p_a = atmospheric pressure constant, J_1 is the first invariant of the stress tensor, R is the term related to the cohesive (tensile) strength, $\bar{c}$, Fig. 1a, $S_r = \frac{\sqrt{27}}{2}\frac{J_{3D}}{J_{2D}^{1.5}}$, J_{3D} is the third invariant of the deviatoric stress tensor, n is the parameter related to the transition from compressive to dilative volume change, γ and β are the parameters associated with the ultimate surface, Fig. 1(a) and α is the yielding or hardening or growth function; in a simple form, it is given by

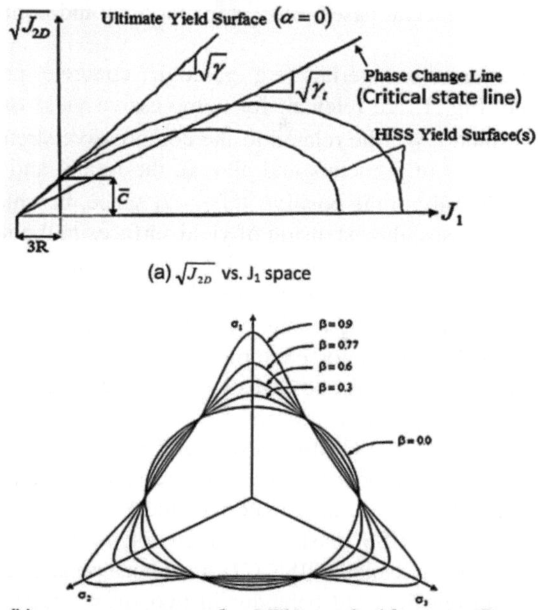

(b) $\sigma_1 - \sigma_2 - \sigma_3$ space; $\beta < 0.756$ required for convexity

Fig. 1. Hierarchical Single Surface (HISS) yield surfaces in two stress spaces

$$\alpha = \frac{a_1}{\xi^{\eta_1}} \tag{2}$$

where a_1 and η_1 are the hardening parameters, and ξ is the accumulated or trajectory of plastic strains, given by

$$\xi = \xi_v + \xi_D \tag{3}$$

Here the accumulated volumetric plastic strain is given by

$$\xi_v = \frac{1}{\sqrt{3}} |\varepsilon_{ii}^p| \tag{4a}$$

and accumulated deviatoric plastic strain is given by

$$\xi_D = \int \left(E_{ij}^p E_{ij}^p \right)^{1/2} \tag{4b}$$

where ε_{ii}^p is the plastic volumetric strain, and E_{ij}^p is the plastic shear strain tensor. In the HISS model, the yield surface grows continuously and approaches the *ultimate* yield,

Fig. 1; it can include, as special cases, other conventional and continuous yield plasticity models.

For compression intensive materials (e.g. geologic, concrete, powders) the model and the yield surface, Fig. 1, are relevant for compressive yield only in the positive $\sqrt{J_{2D}}$ - J_1 space, in which $\bar{c}$ will be related to the compressive strength. Similarly, for tension intensive materials (e.g. metals and alloys), the model and yield surfaces are relevant for tensile yield only in the positive $\sqrt{J_{2D}}$ - J_1 space, in which $\bar{c}$ would denote tensile strength. In both cases, the extension of yield surfaces in the negative J_1 - axis is not usually relevant; they are shown essentially for convenience of plotting. Sometimes, the extended yield surfaces in the negative J_1 – axis have been used with an *ad hoc* model for materials experiencing tensile conditions, which may not be realistic. As discussed later (HISS-CT model), for example, when a material experiences tensile stress (during deformation), it would be realistic to use the model (e.g. HISS) defined on the basis of tensile tests, and *vice versa*.

HISS-CT Plasticity Model for Compressive and Tensile Behavior: It is difficult to develop the same constitutive model for both compressive and tensile yield behavior. However, the yielding behavior of many materials under compression and tension can be modeled using the same HISS plasticity framework; such models that allows for both compression and tension, called, HISS-CT, have been presented [15, 16]. Figure 2 shows compressive and tensile yield behavior in two stress spaces. Here, for convenience, both tensile and compressive yielding is shown on the same stress space. The parameters for the two models are determined from separate compressive and tensile tests.

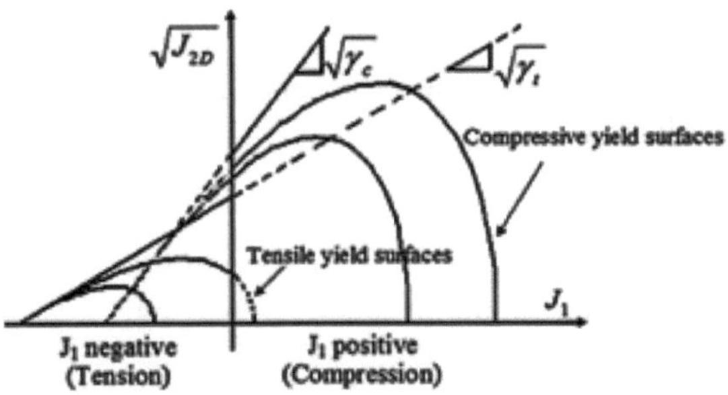

Fig. 2. HISS surfaces for compressive and tensile yield

Comments: The HISS plasticity model allows for continuous yielding, volume change (dilation) before peak, stress path dependent strength, effect of both volumetric and deviatoric strains on the yield behavior, but it does not allow for discontinuities. The HISS surface, Eq. (1), represents a unified plastic yield surface, and most of the previous conventional and continuous yield surfaces can be derived as special cases [1]. Also, the HISS model can be used for nonassociative and anisotropic hardening responses, etc.

Creep Behavior: Many materials exhibit creep behavior, increasing deformations under constant stress or stress relaxation under constant strain (displacement). A number of models have been proposed for various types of creep behavior, viscoelastic (ve), viscoelasticplastic (vep) and viscoelasticviscoplastic (vevp); they are also based on the assumption of continuum material. A generalized creep model has been proposed under the disturbed state concept (DSC) [1]. It is called Multicomponent DSC (MDSC) which includes ve, vep and vevp versions as special cases. Figure 3 shows rheological models for different versions of the MDSC. The equations governing these models, parameters and their determination are given in [1].

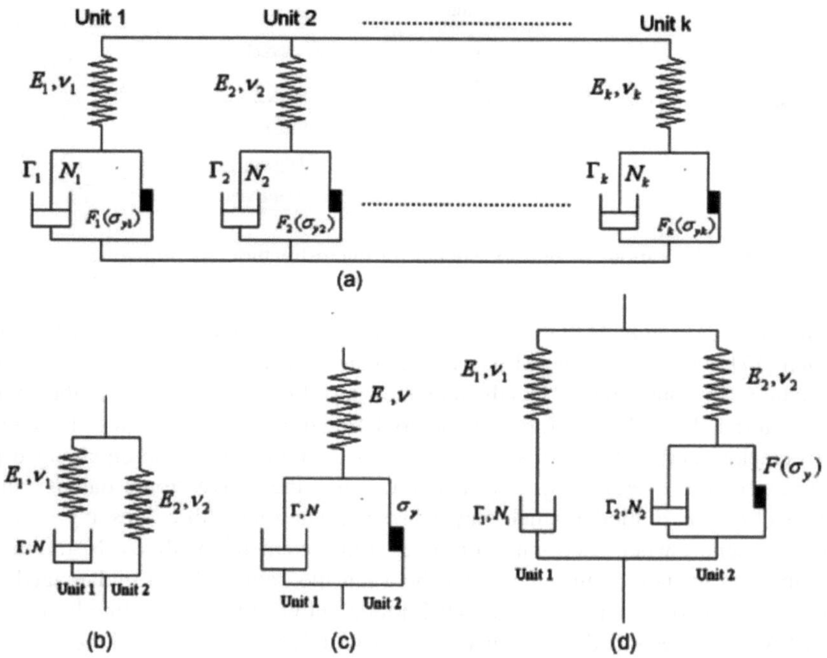

Fig. 3. Rheological versions for creep behavior: (a) MDSC models, (b) viscoelastic (ve), (c) elasticviscoplastic (evp), (d) viscoelasticviscoplastic (vevp)

2.2 Discontinuous Materials

Models based on theories of elasticity, plasticity and creep assume that the material is initially continuous and remains continuous during deformation. However, it is realized that many materials contain discontinuities (microcracks, dislocations, etc.), initially and during loading. During deformation, they coalesce and grow, and separate, resulting in microcracks and fractures, with consequent failure. Since the stress at a point implies continuity of the material, theories of continuum mechanics may not be valid for such discontinuous materials.

There are a number of available models to account for such discontinuities. Chief among those are considered to be fracture mechanics, damage mechanics, micromechanics, microcrack interaction, gradient and Cosserat theories [1, 12]. Most of them combine the effect of discontinuities and microcracks, with the continuum behavior; they are depicted in Fig. 4. Most of such models account for discontinuities by introducing external enrichments or enhancements into models such as the continuum damage [12]. In contrast, the DSC allows for the effects implicitly and does not require external enrichments.

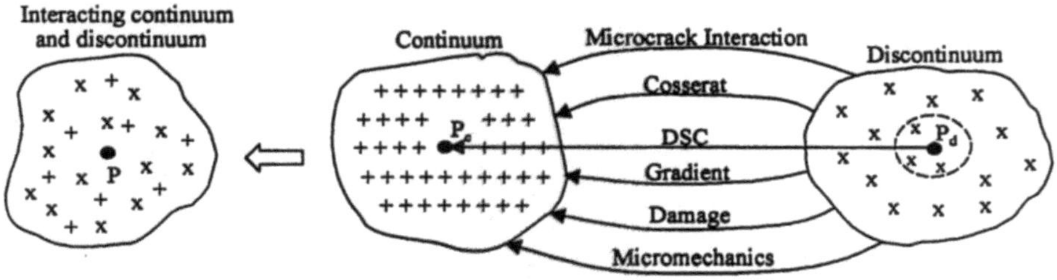

Fig. 4. Various models to account for microstructural discontinuities

DSC and Damage Models: A major limitation of the conventional damage approach [17] is that the damaged part of the material is assumed to have *no strength*, i.e. it cannot carry any stress. Thus the damaged part does not contribute to the strength of the material. In reality, before the material fractures completely, the damaged or (micro) cracked part does possess certain strength, interacts with the undamaged part, and influences the overall observed behavior. Thus, the conventional damage model could result in a *local* model which may cause computational difficulties, e.g., spurious mesh dependence, when used with computer (finite element) methods. Note that it is very important to include the interaction between the damaged and undamaged parts for developing a consistent model; the DSC model allows for the coupling between the RI and FA parts, described later. There is a basic difference between the damage and DSC models; the former is based on physical damage (cracking), while the DSC is based on the behavior of a material considered as a mixture of RI (continuum) and fully adjusted (FA) parts, whose behavior are coupled to yield the observed response [1]. Furthermore, the DSC can be used to characterize stiffening or healing in which the disturbance can decrease [2, 3].

Enrichment of Conventional (Damage), Microcrack Interaction Model: Various schemes have been proposed based on the enrichment of conventional damage model and microcrack interaction in order to account for the coupling between the behaviors of the undamaged and damaged parts in a material element [12]. Various enrichments which are combined with the continuum response can include such models as the Cosserat and gradient plasticity theories, and microcrack interaction. In the microcrack interaction model [18], the weighted effect (strength) of a collection of microcracks is combined with the behavior of the continuum part. Thus it allows for the interaction between damaged and undamaged parts and can provide a n*onlocal* model. However, the definition of a collection of cracks and implementation in computer methods may not be straight forward. Comments on other models that account for discontinuities are given in the foregoing description.

3 Disturbed State Concept (Dsc)

The DSC is a general and unified approach that can accommodate most of the forgoing factors including the effect of discontinuities and provide a hierarchical framework that can include many of the above models as special cases [1–4, 19–26]. One of the important attributes of the DSC is that its mathematical framework can be specialized for contacts (interfaces and joints), thereby providing consistency in using the same model for both solids and contacts [1].

The origin of the DSC constitutive modeling can be traced to the papers by Desai [19, 20] on the subject of behavior of overconsolidated soils and free surface flow in porous materials. The DSC is based on rather a simple idea that the behavior of a deforming material, considered as a mixture of components of various material states, can be expressed in terms of the behaviors of the components. Thus, the behavior of a dry material can be defined in terms of the behavior of material in continuum (called relative intact-RI) state, and microcraked part which approaches, in the limit, to the fully adjusted (FA) state. The RI part can experience microcracking due to self-adjustment or self-organization, which can approach an asymptotic FA state like the critical state, which can be considered as *constrained liquid-solid* state and defined accordingly [1], or *constrained liquid* state defined by using the bulk modulus. It is called the Fully Adjusted (FA) state because in the limit the RI material experiences transition to the FA state, which can denote full failure. Figure 5 shows the RI and FA responses and disturbance in the DSC.

The behavior of the FA part is unattainable (or unmanifested) in practice because it cannot be measured; therefore, a state, somewhere near the residual or ultimate, Fig. 5 (c) can be chosen as approximate FA state. The space between the RI and FA denoted by (i) and (c), respectively, can be called the domain of deformation, whose observed or average behavior (can be called manifested since it can be measured) occurs between the RI and FA states, Figs. 6 and 7. The deviation of the observed state from the RI (or FA) states is called *disturbance,* and is denoted by D. It represents the difference between the RI and observed behavior or difference between the observed and FA behavior.

By defining the observed material behavior in terms of RI (continuum) and fully adjusted parts, the DSC provides for the *interaction* between two parts of the material behavior rather than on the behavior of particle(s) at the micro level. Thus, the emphasis is on the modeling of the *collective behavior of interacting mechanism in clusters of* RI and FA states thereby yielding a model that is considered to be holistic. These comments are similar to those in the self-organized criticality concept [27], which is used to simulate catastrophic events such as avalanches and earthquakes. In this context, the DSC assumes that as the loading (deformation) progresses, the

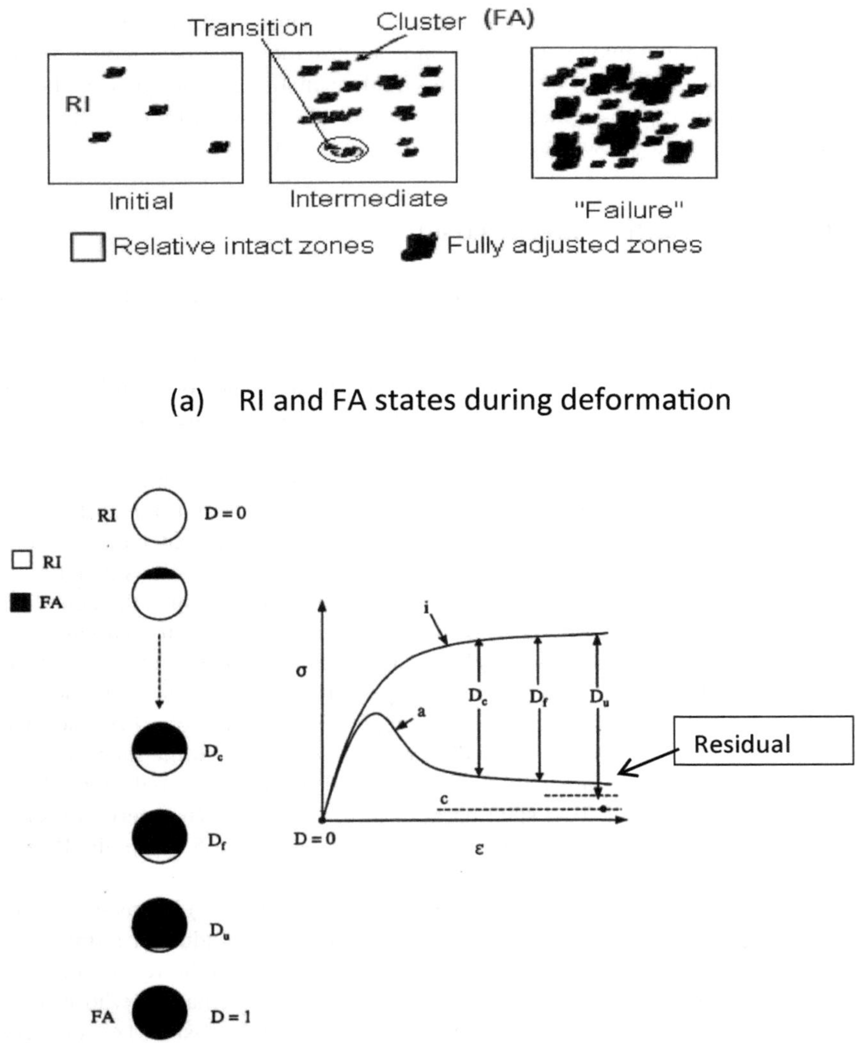

(a) RI and FA states during deformation

(b) Symbolic representation (c) Schematic of stress -strain behavior in DSC

Fig. 5. RI and FA sates and Disturbance in DSC

material in the continuum state transforms continuously into the FA state through modification in the microstructure of the material. The definition of the DSC is not based on the behavior at the microlevel (say, as in micromechanics); rather it is based

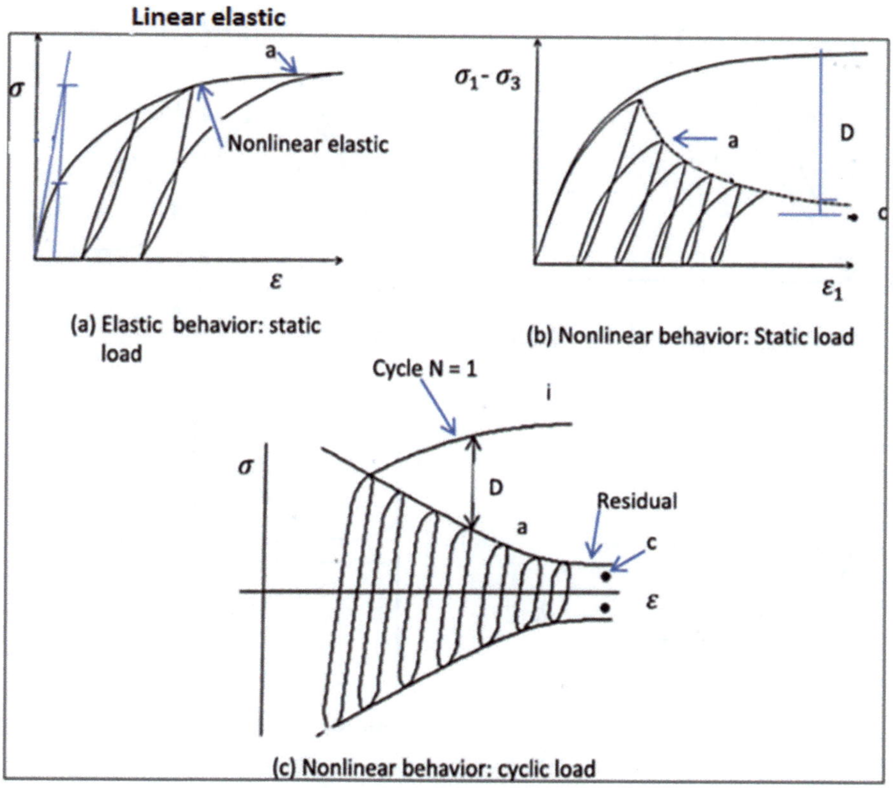

Fig. 6. Schematics of RI (i), FA (c) and disturbance (D)

on the definition of the behavior of the material clusters in the RI and FA states defined from the measured behavior in those states, Fig. 5a.

The behaviors of the RI and FA can be defined from laboratory or field tests, and then the observed behavior is expressed in terms of the behaviors of the RI and FA parts. Assume that the material is continuous from the beginning and remains so during deformation. Such a behavior is called that of RI State, which contains no disturbance. As stated before, the fully adjusted behavior is related to the strength of the material in the FA state. Some of the ways to define RI and FA responses are given below.

Figure 6a shows the continuum response as linear elastic, which can be considered as the RI state. However, the observed response can be nonlinear (elastic), due to factors such as cracking. The FA response can be assumed to have a (small) finite strength. The disturbance can be defined as the difference between linear elastic and nonlinear elastic responses. Figure 6b shows a strain softening response. Here the RI response can be assumed to be nonlinear elasticplastic (e.g. HISS model) and the FA response based on the critical state concept. Figure 6c shows cyclic response. Here the RI response can be adopted as the extended response in the first cycle. The FA response can be assumed to be asymptotic as it becomes steady (c) after a number of cycles.

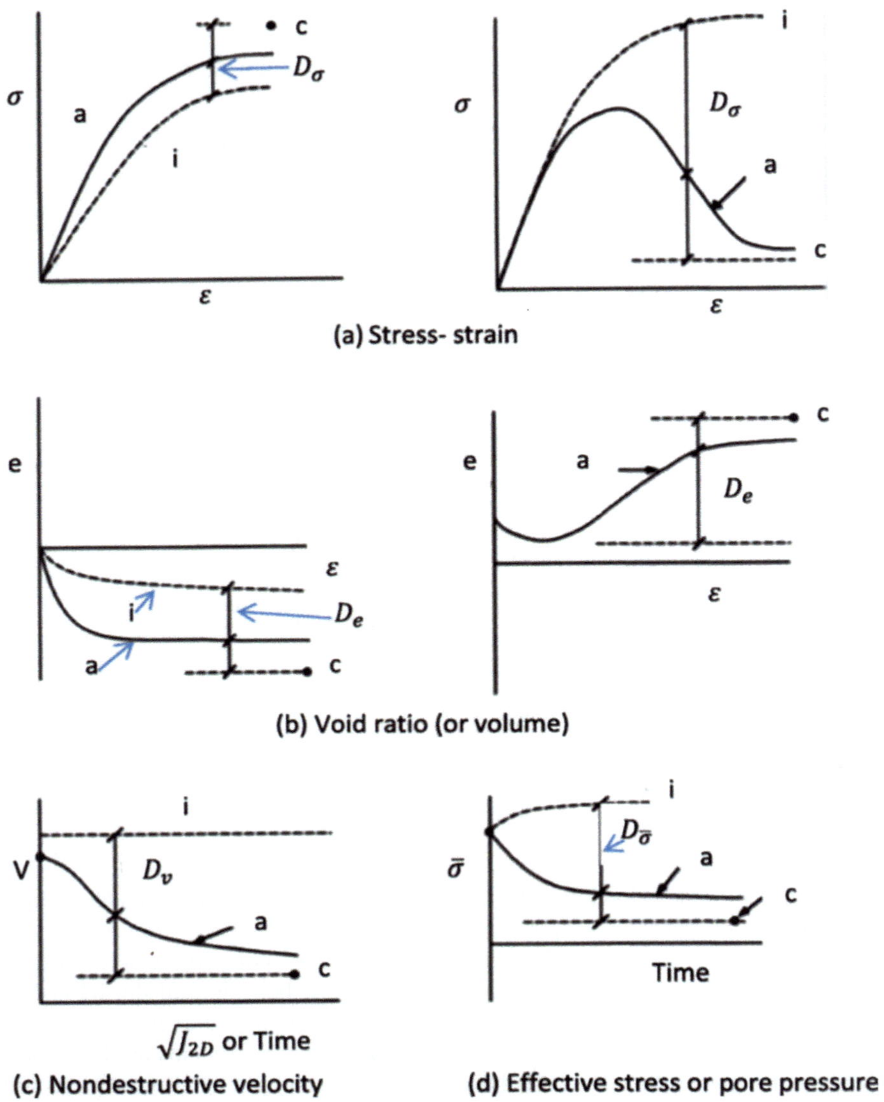

Fig. 7. Various tests behavior to define disturbance

RI State: In many cases, the RI behavior can be assumed to be linear elastic defined by the initial slope. However, if the material behavior is nonlinear and involves effect of other factors such as coupled volume change behavior, e.g. volumetric change under shear loading, such an assumption will not be realistic. Hence, very often, continuous yield or HISS plasticity is adopted to model the RI response. It can be defined by extending the initial part of the observed curve, Fig. 6b, or by integrating the incremental (plasticity) constitutive equations after finding the parameters based on the initial part of the response, before the peak stress [1].

FA State: An easy way is to assume that the material in the FA state has no strength, just like in the classical damage model [17]; this assumption ignores interaction between RI and FA states, may lead to *local* models, and may cause computational difficulties. The other assumption is to consider that the material in the FA state can carry hydrostatic stress like a *constrained liquid*, in which case the bulk modulus (K) can be used to define the response of the FA state; this is similar to the assumption of plastic flow after yield in classical plasticity. The FA material can be considered as of *liquid-solid* like in the critical state [1, 6], when after continuous yield, the material approaches a state at which there is no change in volume or density or void ratio under increasing shear stress. For fluid saturated materials with drainage with time, the RI behavior can be assumed to be that at time near zero time, and the FA response can be assumed like that of a constrained liquid. A description of the DSC for partially saturated materials is given in [1].

Disturbance: As stated before, disturbance defines the coupling between the RI and FA states, and is represented by the deviation (disturbance) of the observed behavior from the RI state. In general, D can be expressed as

$$\mathrm{D} = \mathrm{D}[\xi, w, S, \phi, \mathrm{t(N)}, \mathrm{T}, \alpha_i] \tag{5}$$

where ξ and w denote internal variables such a accumulated plastic strain and (dissipated) energy, respectively, S is entropy (disorder), ϕ is free energy, t is time, N denotes number of loading cycles, T is temperature and $\alpha_i(\mathrm{I} = 1, 2, \ldots..)$ denote factors like environmental effects, dislocations, and impurities. Various representations for disturbance from measurements including nondestructive data, entropy and dislocation are given in [1, 4, 23].

Disturbance can be determined based on the stress-strain behavior, Fig. 7(a). It can be determined from other tests like void ratio (specific density) vs. strain, Fig. 7(b), nondestructive behavior such a for P- and S- waves velocities, Fig. 7(c), fluid (pore) water pressure or effective stress ($\overline{\sigma}$) vs. time or number of cycles, Fig. 7(d), entropy and dislocation [1, 4, 23]. Figure 8 shows the schematic of the disturbance (D) as function of ξ_D or number of cycles (N) or time (t) as affected by appropriate factors; here D_c, D_f and D_u denote initiation of microcracking or fracture, failure and ultimate disturbance, respectively.

Identification Cracking: For instance, D_c, can be used to identify initiation and growth of microcracking (or fracture) at any locations in material, based on its value determined from laboratory tests, at various incremental steps in the finite element procedure [1, 22, 24]. Thus, as in the fracture mechanics, it is not necessary in the DSC to adopt *a priori* locations of cracks.

The disturbance can defined in two ways, (1) from measurements, Fig. 7, as stated before, and (2) by mathematical expression in terms of internal variables such as (ξ_D):

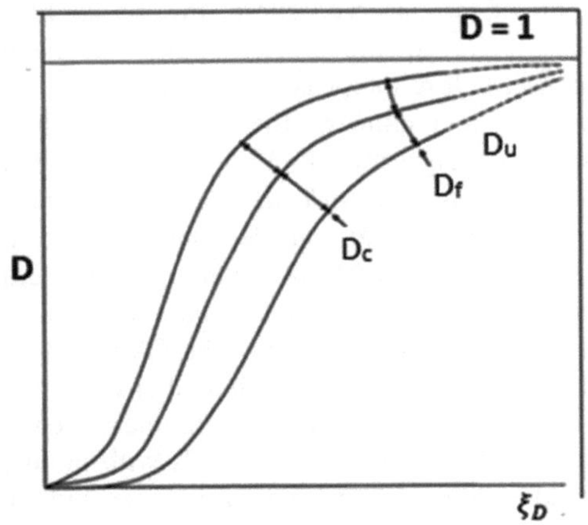

Fig. 8. Disturbance vs. ξ_D (or time or number of cycles) under required factors

3.1 From Measurements, for Example

$$\text{Stress} - \text{strain}: \quad D_\sigma = \frac{\sigma^i - \sigma^a}{\sigma^i - \sigma^c} \tag{6a}$$

$$\text{Fluid pressure}: \quad D_p = \frac{p^i - p^a}{p^i - p^c} \tag{6b}$$

$$\text{Nondestructive velocity}: \quad D_v = \frac{V^i - V^a}{V^i - V^c} \tag{6c}$$

where σ^a is the measured stress, p^a is the measured fluid pressure, and V^a is the measured nondestructive velocity, and i and c represent RI and FA responses. Disturbance can be expressed in terms of free energy, disorder, and dislocations [1, 4, 23].

3.2 Mathematical Expression for D

An equation for disturbance, D, can be expressed using the (Weibull) function in terms of internal variable such as accumulated (deviatoric) plastic strains (ξ_D) or plastic work:

$$D = D_u\left[1 - \exp\left(-A\xi_D^Z\right)\right] \tag{7}$$

where A, Z and D_u are the parameters. The value of Du is obtained from the ultimate FA state, Fig. 6b. Equations 6a, 6b and 6c can be used to find the disturbance, Fig. 7, at various points on the response curves, which are substituted in Eq. 7 to find the parameters, A and Z.

Note that the expression in Eq. (7) is similar to that used in various areas such as biology to simulate birth to death or growth and decay, and in engineering to define damage in classical damage mechanics, and disturbance in the DSC. However, the concept of disturbance is much different from damage; disturbance defines deviation of observed response from RI (or FA) states, in the material treated as a mixture of interacting components, while damage represents physical damage or cracks.

3.3 DSC Equations

Once the RI and FA states and disturbance are defined, the incremental DSC equations based on equilibrium of a material element can be derived as [1]:

$$d\sigma_{ij}^{a} = (1-D)d\sigma_{ij}^{i} + Dd\sigma_{ij}^{c} + dD\left(\sigma_{ij}^{c} - \sigma_{ij}^{i}\right)$$
$$or \qquad (8)$$
$$d\sigma_{ij}^{a} = (1-D)C_{ijkl}^{i}d\varepsilon_{kl}^{i} + DC_{ijkl}^{c}d\varepsilon_{kl}^{c} + dD\left(\sigma_{ij}^{c} - \sigma_{ij}^{i}\right)$$

where σ_{ij} and ε_{ij} denote stress and strain tensors, respectively, C_{ijkl} is the constitutive tensor, and dD is the increment or rate of D. The conventional continuum (elasticity, plasticity, creep) models can be derived as special cases by setting D = 0 in Eq. (8). If $D \neq 0$, the equations account for microstructural modifications in the material leading to fracture, failure, and instability like failure and liquefaction (in saturated materials), and stiffening or healing. The latter can be defined corresponding to the critical disturbance, D_c, Fig. 8, obtained from measured response of the material.

3.4 Parameters

The basic DSC model contains the following parameters:

Relative Intact (RI): Elastic: Young's modulus, E, and Poisson's ratio, v , (or shear modulus, G and bulk modulus K), and *Plasticity: e.g.* Mohr Coulomb: cohesion, c and angle of internal friction, φ, or (c) HISS plasticity: ultimate yield, γ and β; phase change (transition from compaction to dilation), n; continuous yielding, a_1 and η_1, and cohesive strength intercept, $\bar{c}$ (R).

Fully Adjusted (FA): The response of the material in the FA state can be defined in a number of ways:(i) Hydrostatic strength as *constrained liquid* with bulk modulus, K; or (ii) Constrained *liquid-solid* as in the critical state: slope of the critical state line, $\bar{m}$, slope of void ratio vs. mean pressure, λ, initial void ratio, e_o [1, 6]. The former may be appropriate for metals and alloys, and latter for geologic materials like soils, rocks, concrete, and powders. It is not appropriate to assume the material at the FA state like void with zero strength, as in the classical damage mechanics because such an assumption ignores interaction between materials in RI and FA states.

Disturbance: The parameter D_u can be obtained from Figs. 5c, 6b; often, a value near unity can be used. Parameters A and Z are obtained by first determining various values of D from the test data by using Eqs. 6a, 6b, 6c, and then plotting logarithmic form of Eq. 7.

Comments: Most of the above parameters in the DSC have physical meanings, i.e. almost all are related to specific states in the material response, e.g. elastic modulus to the unloading slope of stress-strain behavior, Fig. 6, γ and β to the ultimate state, Fig. 1 and n to the transition from compactive to dilative volume change. Their number is equal or lower than that of previously available models of comparable capabilities. They can be determined from standard laboratory tests such as uniaxial, shear, triaxial and/or multiaxial. The procedures for the determination of the parameters are provided in various publications [1–3, 14].

The hierarchical property in the DSC-HISS model allows adoption of parameters only for the factors needed for the behavior of a specific material, e.g. if yielding and disturbance are not involved, only the elastic parameters are needed, and so on.

4 Contacts: Interfaces and Joints

Behavior at contacts or interfaces between two (different) materials plays a significant role in the overall response of an engineering system [1, 28–31]. One of the main advantages of the DSC is that its mathematical framework can be specialized for contacts (interfaces and joints). Hence, use of the same framework in the DSC for both solids and contacts provides required consistency. This is in contrast to the common use of different models for solids and contacts in the same problem, e.g. a plasticity model for solids and bilinear elastic model for interfaces. A brief description of the DSC for interfaces is given below. Figure 9 shows schematics of two- and three-dimensional contacts.

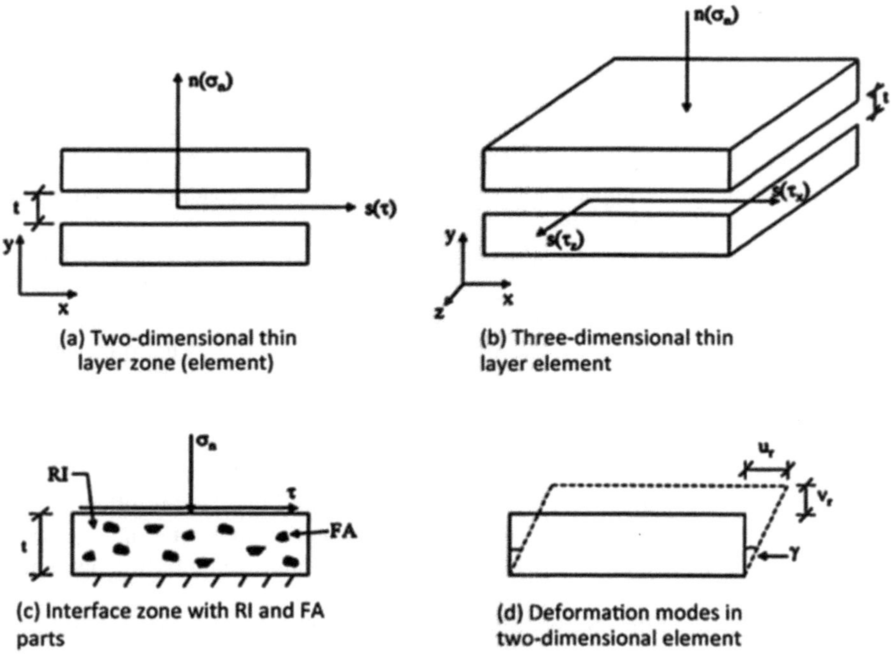

Fig. 9. Two- and three-dimensional contacts

Relative Intact (RI) and FA Responses: As described in [1, 28–31] the RI and FA responses can be defined based on appropriate interface (normal and shear) tests under different normal stresses, roughness, (initial) pore water pressure, amplitudes of cyclic loadings, etc.

The yield function specialized from Eq. (1) for two-dimensional interface is given by

$$F = \tau^2 + \alpha\sigma_n^n - \gamma\sigma_n^q = 0 \tag{9}$$

where τ is the shear stress, σ_n is the normal stress, which can be modified as σ_n + R, R is the intercept along σ_n axis, γ is the slope of ultimate response, n = phase change parameter, which designates transition from compressive to dilative response, q = governs the slope of the ultimate envelope (if the ultimate envelope is linear, q = 2), and α is the growth or yield function given by

$$\alpha = \frac{h_1}{\xi^{h_2}} \tag{10}$$

where h_1 and h_2 are hardening parameters, and ξ is the trajectory of plastic relative horizontal (u_r) and vertical (normal) (v_r) displacements, respectively, given by

$$\xi = \int \left(du_r^p \cdot du_r^p + dv_r^p dv_r^p\right)^{1/2} = \xi_D + \xi_V \tag{11}$$

ξ_D and ξ_V denote accumulated plastic shear and normal displacements, respectively, and p denotes plastic.

Disturbance Function: Disturbance can be derived from measured stress-strain, volumetric, effective stress (or pore water pressure) and cyclic behavior of the contact as in the case of solids, Fig. 7. It can be expressed as in Eq. 7 using ξ in Eq. 11.

ANALYSIS of DSC: For reliable and consistent predictions, a constitutive model must satisfy certain mathematical characteristics such as localization and mesh dependence. The DSC is based on a weighted definition of the stress, and interaction between the RI and FA parts. As a result, it can provide freedom from spurious mesh dependence, and consistent localization; details of these aspects are presented in [1–3, 32, 33].

4.1 Comments

The DSC has been published in a number of publications [1–4]; these works include application of the DSC by the author and coworkers, and by others for materials such as soils, structured soil, masonry, concrete, asphalt concrete, fully and partially saturated materials, rock and rockfills, pavement materials, metals, alloys, ceramics, polymers and silicon, and contacts such as interfaces and joints.

Predictions from finite element procedures with the DSC have been validated with respects of specimen and boundary value problem levels with measurements in laboratory and field for a wide range of problems in Geomechanics/geotechnical engineering, structural mechanics, earthquake analysis including liquefaction as a microstructural instability, rock mechanics, glacial mechanics, and electronic packaging (composites). It has been used for applications beyond material behavior, e.g., developing expressions for earth pressures, computation of pile capacity and free surface fluid flow.

4.2 Validations and Applications

Typical applications and validations of the DSC model for geologic materials and interfaces (emphasized here) and an example of practical boundary value problem are presented below.

4.3 Lime-Stabilized Collapsible Loess

Testing and constitutive modeling of lime stabilized unsaturated loess has been reported in [34]. The results obtained from filter paper and unsaturated odometer tests on a lime treated loessial soil have been analyzed by considering the effective stress approach with an empirical model for explaining the load-collapse behavior of the soil. Using the disturbed state concept (DSC), a coupled semi-empirical hydro -mechanical model is developed to predict the disturbance level and calculate the soil strain due to application of the vertical stress in the K_o condition. Both models presented in this research are introduced as functions of two important governing state and material variables, namely matric suction and lime content predictions.

The soil response in any condition, P_a, can be expressed in terms of reference states P_{RI} and $P_{FA,}$ relative intact and fully adjusted, respectively:

$$P_a = P_{RI} - (P_{RI} - P_{FA}) \times D \tag{12}$$

where the parameter D is the disturbance function. Commonly, D can be defined as in Eq. 7, in the form of an exponential function based on appropriate material response variables such as accumulated deviator plastic strain, ξ_D, and some fitting parameters as presented by (Fig. 10)

According to the SWRC of the tested specimens, a semi-empirical relationship for predicting the values of D_u as a function of matric suction, ψ, and degree of saturation, S_r, in the wetting path is proposed:

$$D_u(\psi) = 1 - \left(\frac{S_{r\_RI}}{S_r(\psi)} \times \frac{\log(1+\psi)}{\log(1+\psi_{RI})} \right)^k \tag{13}$$

where S_{r_RI} and ψ_{RI} (kPa) are degree of saturation and corresponding matric suction at RI state, respectively, and k is a dimensionless parameter whose value depends on the material properties. One feature of the proposed model is that the obtained values of $D_u(\psi)$ for RI and FA states of the soil become zero (the least possible value) and unity (the maximum possible value), respectively.

Comparison is carried out on the nine sets of "Observed data" (*i.e.,* tests conducted under matric suctions of 50, 100 and 200 kPa and three lime contents of 0%, 1%, 3%). Typical predictions are compared with the measured values ($p'_{c,test}$) obtained from unsaturated oedometer tests, Fig. 11. The correlations between predictions and tests data are excellent.

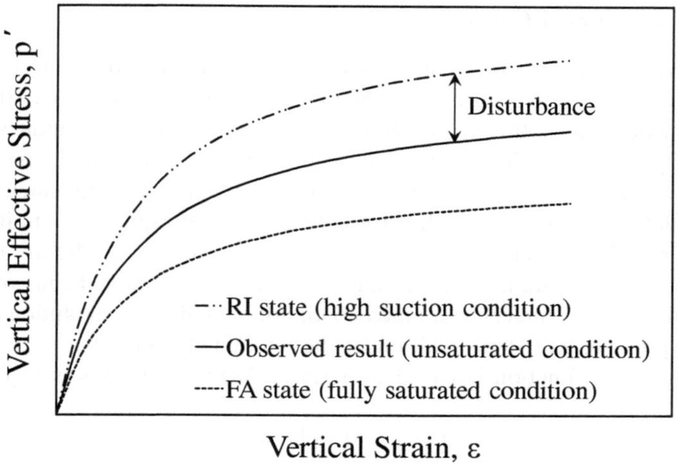

Fig. 10. Schematic of Disturbance (34)

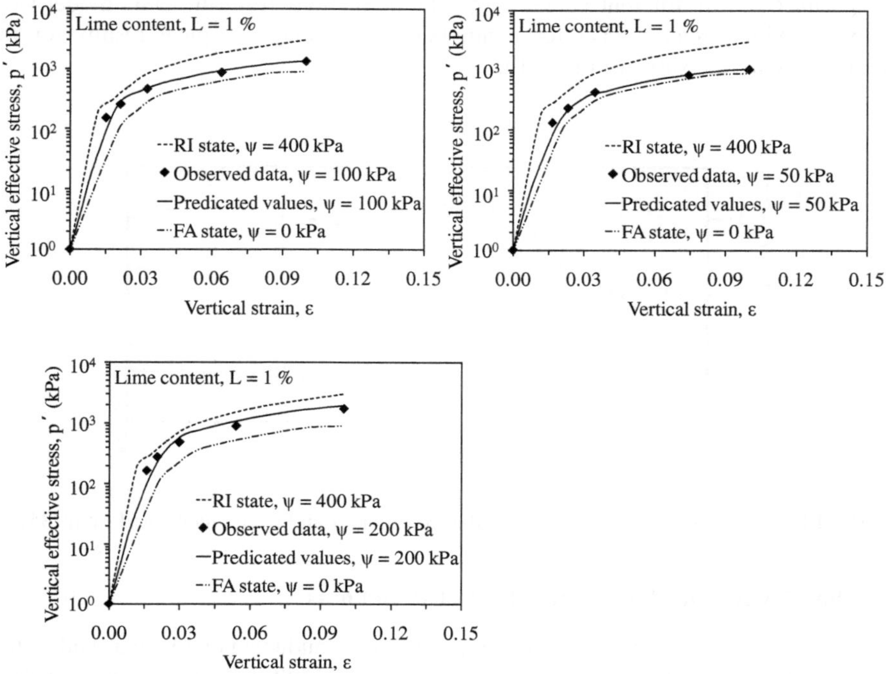

Fig. 11. Comparisons between observed and DSC predictions: Lime content = 1%, $\psi =$ 50, 100, 200 kPa (34)

4.4 Overconsolidated Clays

A constitutive model was proposed for overconsolidated clays by using the disturbed state concept [35]. A new disturbance function was defined for describing the extent of the over consolidation ratio on the strength, dilatancy and deformation of overconsolidated clays, and was incorporated into the potential-failure and dilatancy stress ratios, potential-failure and dilatancy lines, potential-failure and dilatancy surfaces at the current and referenced states. The proposed DSC could capture the strain hardening and volumetric contraction behavior of normally consolidated or slightly overconsolidated clays, and the strain-softening and volumetric expansion behavior of highly overconsolidated clays under the drained condition, and also the evolution of strength, excess pore pressure and stress path under the undrained condition.

The disturbance function D_{OCR} of overconsolidated (OC) clays is expressed as

$$D_{OCR} = 1-(1/O_\delta)^\xi \tag{14}$$

where O_δ is the ratio of potential failure mean effective stress, ξ in a material constant that controls the effect of O_δ on the disturbance function; O_δ Varies during the loading; its initial value is equal to the OC ratio of the clay. Figure 12 shows the variation of D_{OCR} with O_δ with different values of ξ. The predictions from the DSC model were compared with observed data for a number of OC clays. Typical results for the Fujinomori clay are shown in Fig. 13.

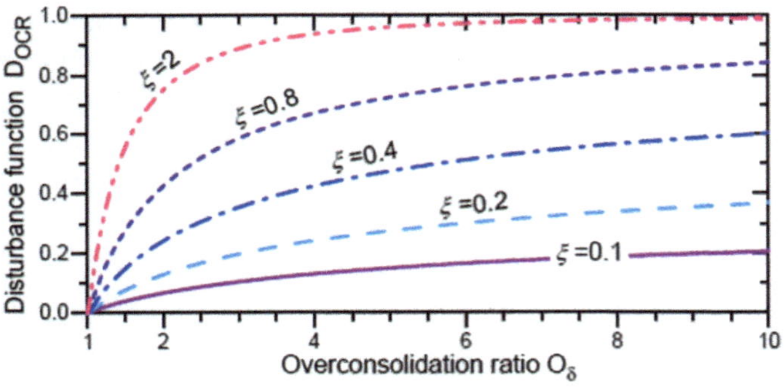

Fig. 12. Relationship between the disturbance function and overconsolidated ratio (35)

4.5 Sand-Steel Interfaces with Effect of Roughness

Monotonic and cyclic simple shear experiments were conducted on Ottawa sand – steel interfaces under undrained conditions using the Cyclic Multi Degree Of Freedom shear device (CYMDOF-P) with pore water pressure measurements [36]. The DSC model was used and the concept of critical disturbance was developed to identify the initiation of liquefaction in saturated Ottawa sand-steel interfaces, where microstructural changes in the deforming interface zone lead to instabilities like liquefaction. A new and unique

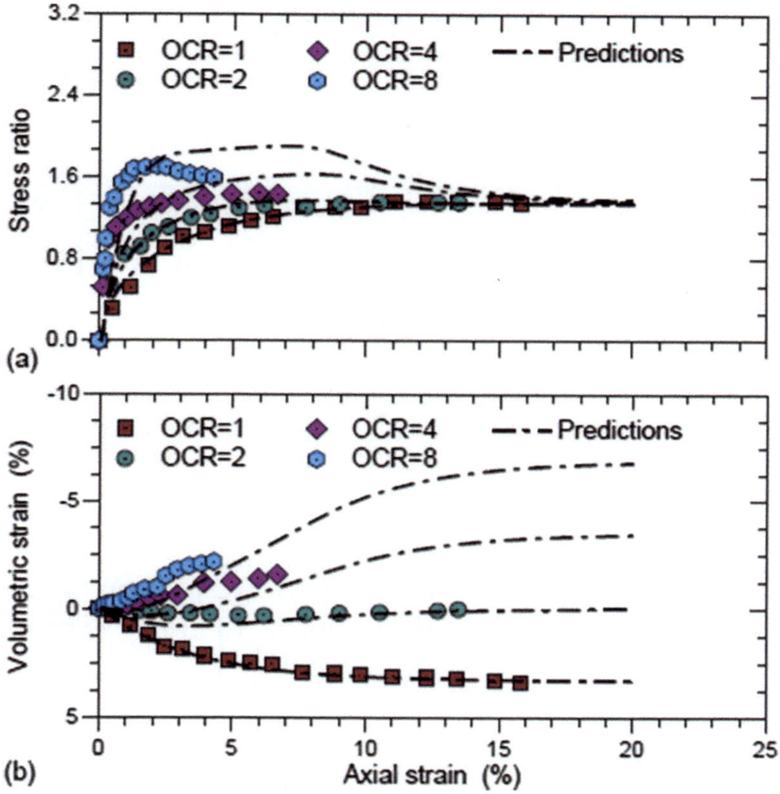

Fig. 13. Comparisons between model predictions and test data of Fujinomori clay: (a) relationship between deviatoric stress and axial strain; (b) relationship between pore water pressure and axial strain (35)

constitutive modeling, testing and validations for interface behavior affected by factors such as surface roughness, confining pressure, amplitude of loading and frequency, are presented [36].

Table 1 shows the DSC parameters expressed as the function of roughness.

Figure 14 shows typical comparisons between DSC predictions and cyclic tests data. Overall, the model is found capable of predicting behavior of static, one-way and two-way cyclic loading.

4.6 Polymer Concrete-Sand Interfaces

Two sets of interface direct shear tests (over 40) were conducted on the polymer concrete and sand interface [37]. Moreover, two sets of tests (over 30) were performed on the cement concrete and sand interface for comparison. Based on the experiments, the shear stress versus tangential displacement curves at different normal stresses, the interface friction angles, and the adhesion were obtained for each interface. Then, three

Table 1. DSC Parameters for Saturated Interfaces (36)

Group	Parameter	Surface roughness			Relations Between DSC Parameters and Surface Roughness R_s	
		RI	RII	RIII	Regression	Correlation factor r^2
Elastic parameters	K_n (kPa/mm)	300	440	476	$K_n = 280.9R_s^{1.442}$	0.958
	K_s (kPa/mm)	101.7	153.3	170.0	$K_s = 94.2R_s^{1.596}$	0.970
Plasticity parameters	γ	0.196	0.377	0.418	$\gamma = 0.173R_s^{2.318}$	0.993
	n	2.98	3.24	3.52	$n = 2.88R_s^{0.491}$	0.976
	h_1	0.030	0.072	0.070	$h_1 = 0.028R_s^{2.74}$	0.853
	h_2	0.304	0.259	0.257	$h_2 = 0.31R_s^{-0.536}$	0.898
Critical state parameters	$\bar{m}$	0.422	0.564	0.597	$\bar{m} = 0.4R_s^{1.085}$	0.956
	λ	0.071	0.050	0.049	$\lambda = 0.074R_s^{-1.182}$	0.903
	e_0^c	0.518	0.536	0.527	–	–
Disturbance parameters	D_u	1.0	1.0	1.0	–	–
	Z	1.178	1.170	1.105	$Z = 1.2R_s^{-0.175}$	0.695
	A	0.0026	0.0067	0.0092	$A = 0.002R_s^{3.891}$	0.985

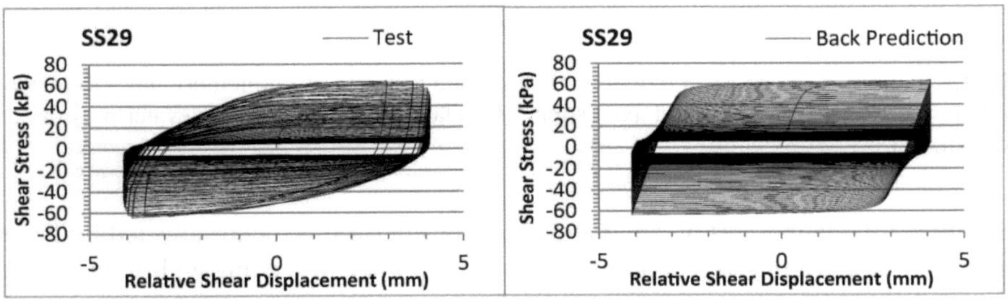

Fig. 14. Observed and Back Predicted Shear Stress vs. Relative Shear Displacement for Two-way Cyclic Loading (σ_n = 150 kPa, Amplitude = 5.0 mm, Freq. = 0.1 Hz, RI) (36)

different constitutive models, DSC/HISS, Liu eta al. (2006) and De Gennaro and Frank (2002) (details of latter two are given in Ref. 37), were proposed to predict the observed response. Finally, after a quantitative comparison of the constitutive models, a pile in tension was modeled using the most accurate model for its interfacial behavior. The results of the proposed (DSC) model were in excellent accordance with experimental data, and the results of pile modeling in tension showed the applicability of the model and obtained parameters. Figure 15 shows the comparisons for interface behavior by using three different models. It can be seen that the DSC/HISS models provides the most accurate predictions.

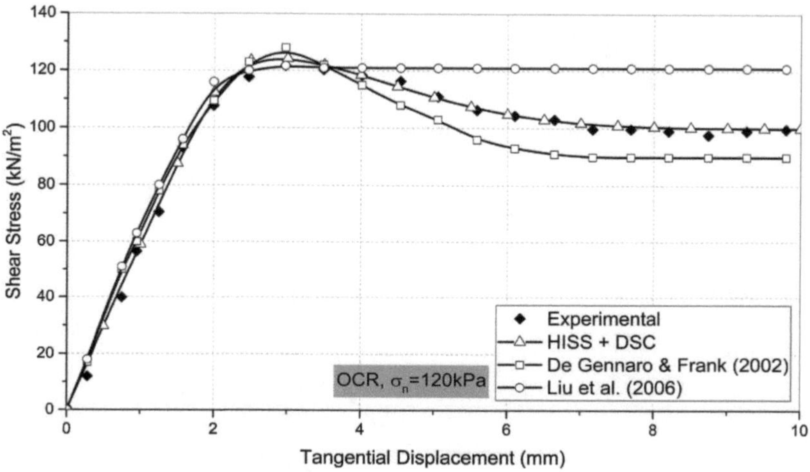

Fig. 15. Comparison between results of different constitutive models (37)

4.7 Softening and Volumetric Behavior of Rocks

A constitutive model was developed and numerical analysis was conducted on various types of rocks under triaxial compression to determine their softening and volumetric responses [38]. Disturbed state concept was utilized and experimental data sets for different types of rocks were used to determine the DSC/HISS material parameters. Verification of the numerical results with experimental data sets showed very good accuracy for various types of rocks under different confining pressures. Typical validations for two rocks, Kirchberg Granite are presented here in Fig. 16.

4.8 Liquefaction Analysis Using the DSC

The disturbance and energy approaches are presented and applied for analysis of liquefaction in the field at Port Island, Kobe, Japan, during the Hyogo-ken Nanbu earthquake, and cyclic laboratory responses of Ottawa and Reid Bedford sands [39]. Correlations between the disturbance and energy approaches are established, and their application potential is identified together with the advantages of the disturbance approach. The field data and liquefaction at the Port Island site have been analyzed and predicted successfully.

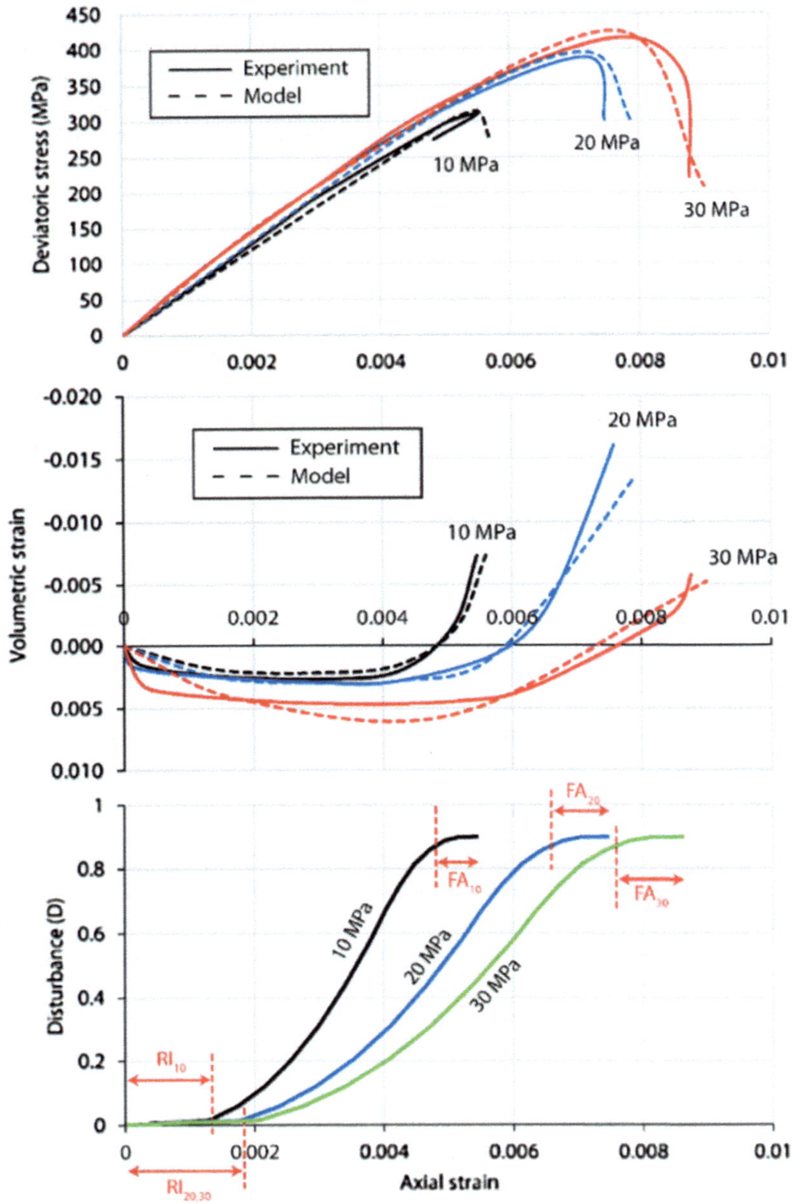

Fig. 16. Experimental and numerical results of deviatoric stress, volumetric strain and disturbance versus axial strain for Kirchberg-II granite (II) (38)

Figure 17 shows the soil profile and the locations of strong motion instruments at different depths at the Port Island site, which were adopted from various publications cited in [39]. Figure 18 shows measured shear wave velocities at the Kobe site. In

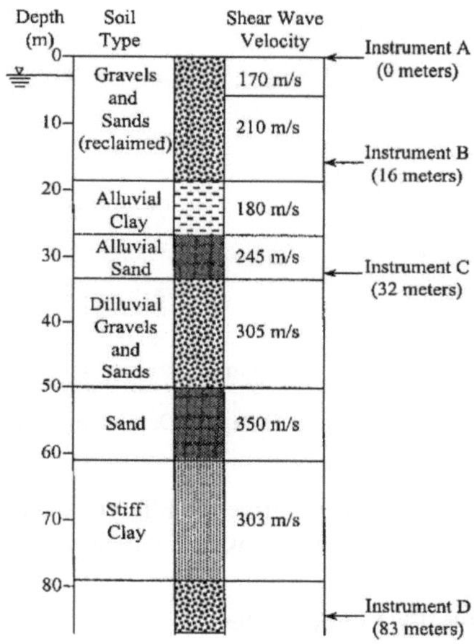

Fig. 17. Details of Soils and Instruments at Port Island, Kobe site (39)

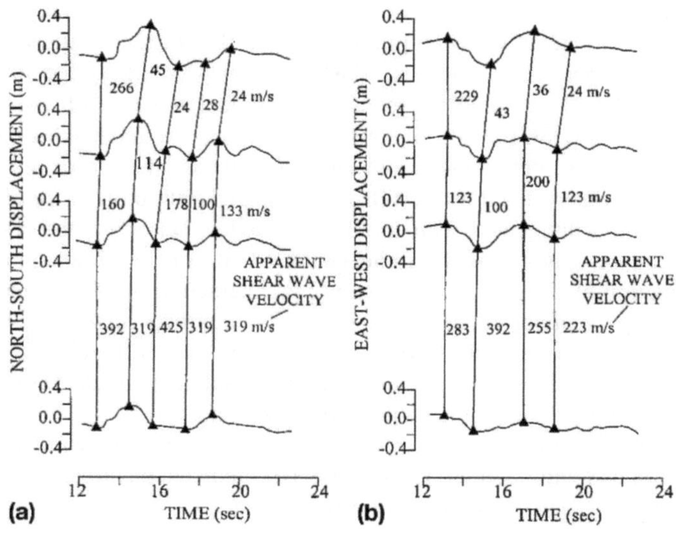

Fig. 18. Measured shear wave velocities with time (Davis and Berrill (1998) (a) North-South; (b) East-West (39)

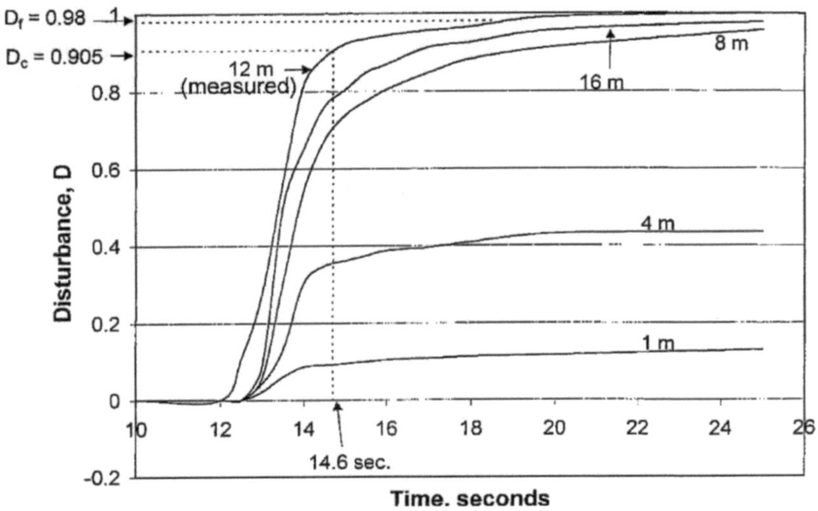

Fig. 19. Disturbance versus Time at Different Depths and time at liquefaction (39)

Fig. 19 are shown the disturbance vs. time at different depths, and time at liquefaction = 14.6 s which compared excellently with the measured value from the field data.

4.9 Dynamics of Laterally Loaded Piles

Analyses and validations of dynamic field piles tests in sand are conducted using nonlinear finite element approach with DSC constitutive model to assess the capability of the models to predict the soil-structure interaction and liquefaction behavior [39]. Instrumented axially loaded pile test at Sabine, Texas (Earth Technology Corporation, 1986) and full-scale laterally loaded pile test at Seal Beach, California (Ertec Western Inc. in Full-scale pile vibration tests, 1981) are analyzed using the finite element program DSC-DYN2D. A modified constitutive model is based on the unified disturbed state concept (DSC) [1]. Both soils and interfaces are modeled using the DSC, and the parameters are determined from laboratory triaxial tests on soils and special shear tests on interfaces. Field test results are compared with those from finite element analyses for both without and with interface behavior. Based on the results of this research, it can be stated that the finite element-DSC model simulation allows for realistic prediction of complex dynamic soil–pile interaction problems, and is capable of characterizing behavior of saturated soils and interfaces involving liquefaction. Typical results only for laterally loaded pile tests are included herein.

Figure 20 and 21 show the finite element mesh and applied lateral displacements, respectively. Figure 22 shows the location of piezometers.

Figure 23 shows observed lateral deflections in comparison with finite element predictions with and without interface. Contours of predicted disturbances for steps 204 and 214 are shown in Figs. 24 and 25, respectively.

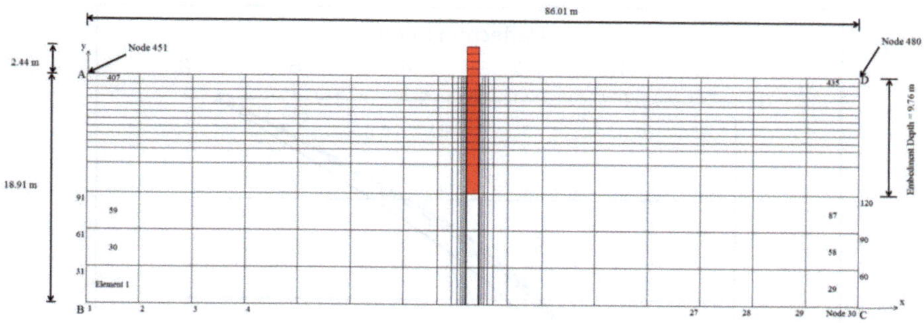

Fig. 20. Finite element mesh (Not to Scale) (39)

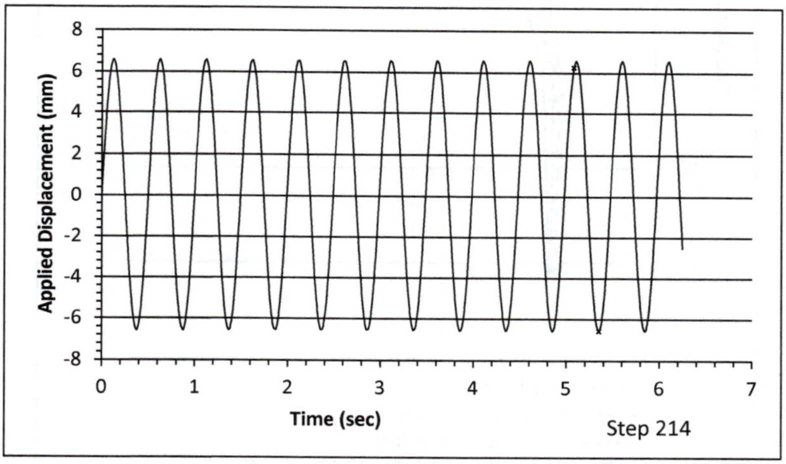

Fig. 21. Applied displacement on laterally loaded pile (39)

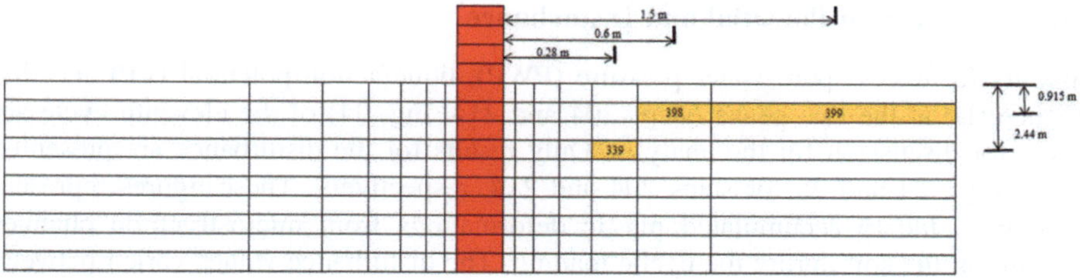

Fig. 22. Elements (339, 398, and 399) corresponding to piezometers locations (Not to Scale) (39)

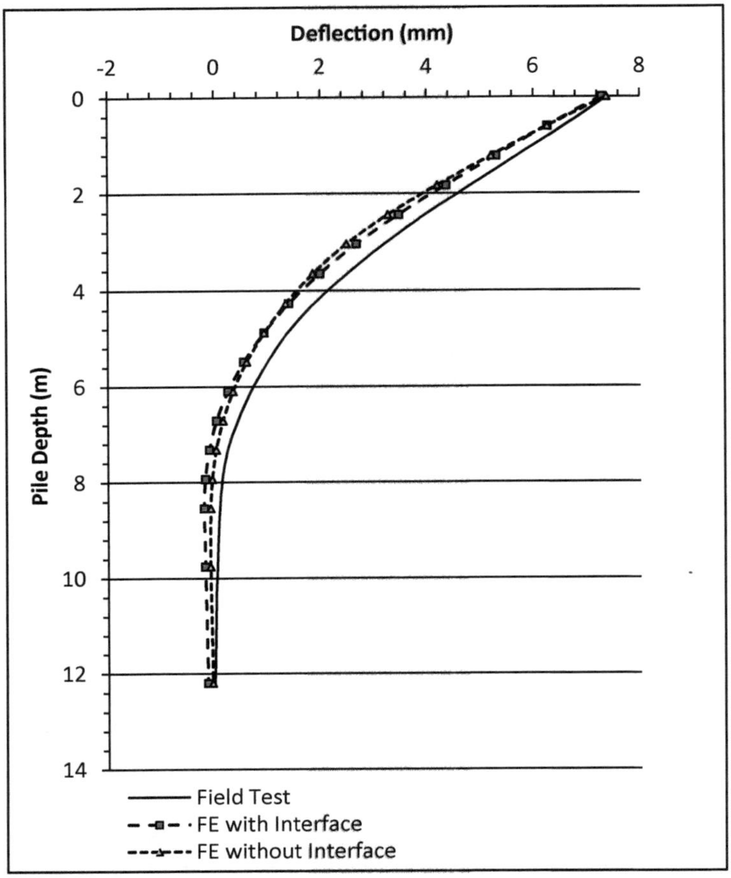

Fig. 23. Effect of interface on pile deflection at step 5 (39)

4.10 Liquefaction Potential and Disturbance

Contours for excess pore water pressure (PWP), liquefaction potential (LP) and disturbance (D) at the two peaks (steps 204 and 214, Fig. 24) of the eleventh cycle are chosen for discussion for the analysis. Only results for the disturbance are presented here in Figs. 24 and 25 for steps 204 and 214, respectively. These figures represent disturbance due to accumulated plastic deformations from microstructural changes occurring in the soil during the cyclic loading. The disturbance values varied between 0.7 and 1.0 immediately around the pile and then reduced and became less than 0.7 in the far regions. Park and Desai (2000) [40] identify the initiation of liquefaction for sand at $D_c = 0.84$. For sand-steel interface, the critical disturbance corresponding to the state of liquefaction, [39], is $D_c = 0.91$. Hence, liquefaction occurs around the pile and in the shallow soil, as reported in the field by Ertec, 1981 [39].

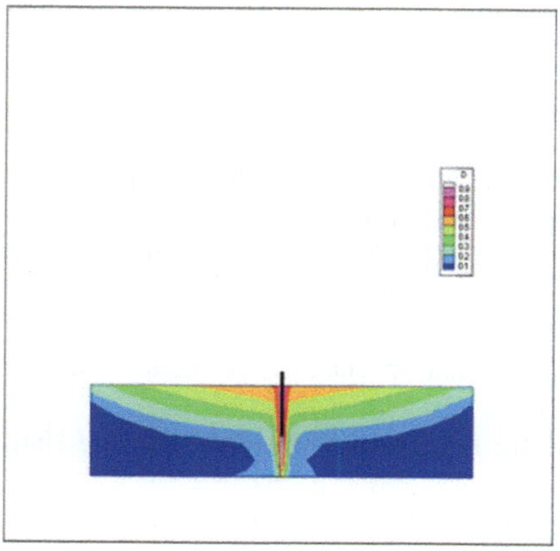

Fig. 24. Disturbance at Step 204 (39)

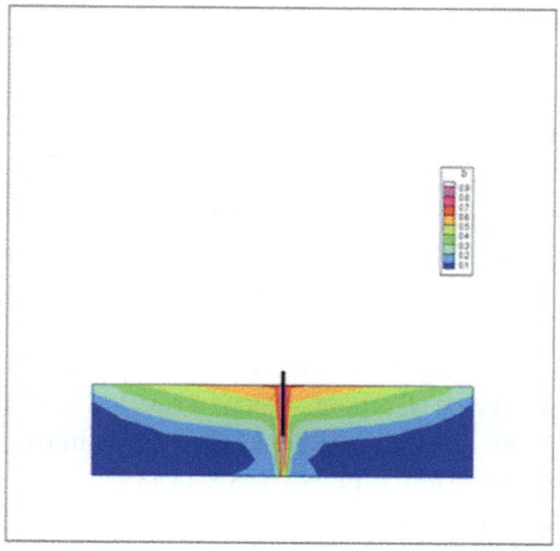

Fig. 25. Disturbance at Step 214 (39)

5 Conclusions

Constitutive modeling of solids and contacts is one of the issues that play a key role in obtaining realistic solutions from conventional and computer procedures. Hence, constitutive modeling is emphasized in this paper. A brief review of a number of available models is presented. Then the disturbed state concept (DSC-HISS) is described, which provides a unified constitutive modeling approach for engineering materials and allows for elastic, plastic and creep strains, stress path dependence, volume change under shear, microcracking and fracturing, softening and degradation, stiffening or healing, all within a single, hierarchical framework. Its capabilities go well beyond other available models yet lead to significant simplifications for practical applications. A major advantage of the DSC-HISS approach is that its formulation for solids can be specialized for interfaces and joints. The DSC-HISS model has been applied successfully for characterization of behavior of many materials such as cohesive and cohesionless soils, rocks, concrete, asphalt, ceramics, metals, alloys, polymers and silicon, and interfaces and joints.

The DSC-HISS model is validated at specimen and boundary value problem levels. Predictions from the finite element procedure with the DSC-HISS model have been compared with measurements from the laboratory and the field for a wide range of problems in Geomechanics/geotechnical engineering, structural mechanics, earthquake analysis including liquefaction as a microstructural instability, rock mechanics, glacial mechanics, and electronic packaging (composites).

References

1. Desai, C.S.: Mechanics of materials and interfaces: the disturbed state concept. CRC Press, Boca Raton, FL, USA (2001)
2. Desai, C.S.: Constitutive modeling of materials and contacts using the disturbed state concept: part 1—background and analysis. Comput. Struct. **146**, 214–233 (2015)
3. Desai, C.S.: Constitutive modeling of materials and contacts using the disturbed state concept: part 2—validations at specimen and boundary value problem levels. Comput. Struct. **146**, 234–251 (2015)
4. Desai, C.S.: Constitutive modeling vital for computational mechanics, expressions. IACM Int. Assoc. Comput. Mech. **41**(17), 12–17 (2017)
5. Desai, C.S., Siriwardane, H.J.: Constitutive Laws for Engineering Materials. Prentice-Hall, Englewood Cliffs, NJ, USA (1984)
6. Roscoe, K.H., Schofield, A., Wroth, C.P.: On yielding of soils. Geotechnique **8**, 22–53 (1958)
7. Mroz, Z.: On the description of anisotropic work hardening. J. Mech. Phys. Solids **15**, 163–175 (1967)
8. Prevost, J.H.: Plasticity theory for soils stress-strain behavior. J. Eng. Mech., ASCE **104**(5), 1177–1194 (1978)
9. Dafalias, Y.F.: A bounding surface plasticity model. Proc., 7th Cong. Appl. Mech., Sherbrooke, Canada, 1979
10. Pastor, M., Zienkiewicz, O.C., Chan, A.H.C.: Generalized plasticity and the modeling of soil behavior, Int. J. Num. Analyt. Comp. Meth. Geomech. **14**(3), 151–190 (1990)
11. Manzanal, D., Pastor, M., Merodo, J.: Generalized plasticity constitutive model basedon state parameters. CMES Comput. Model. Eng. Sci. **55**(3), 17–20 (2010)
12. Mühlhaus, H.B. (ed.): Continuum Models for Materials with Microstructure. Wiley, Chichester, UK (1995)
13. Desai, C.S.: A general basis for yield, failure and potential functions in plasticity. Int. J. Num. Analyt. Meth. Geomech. **4**, 361–375 (1980)
14. Desai, C.S., Somasundaram, S., Frantziskonis, G.: A hierarchical approach for constitutive modeling of geolotic materials. Int. J. Num. Analyt. Meth. Geomech. **10**(3), 225–257 (1986)
15. Desai,C.S.: Unified DSC constitutive model for pavement materials with numerical implementation. Int. J. Geomech. ASCE, **7**(2), 83–101 (2007)
16. Akhaveissy, A.H., Desai, C.S.: Application of DSC model for nonlinear analysis of reinforced concrete frames. Fin. Elem. Anal. Des. J. **50**, 98–107 (2012)
17. Kachanov, L.M.: Introduction to Continuum Damage Mechanics. Martinus Nijhoft Publsihers, Dordrecht, The Netherlands (1986)
18. Bazant, Z.P.: Nonlocal damage theory based on micromechanics of crack interactions. J. Eng. Mech. ASCE **120**, 593–617 (1954)
19. Desai, C.S.: A consistent finite element technique for work-softening behavior. In: Proceedings of the International Conference on Comp. Meth. in Nonlinear Mechanics, J. T. Oden, et al. (editors), Univ. of Texas, Austin, TX, USA, 1974
20. Desai, C.S., Li, G.C.: A residual procedure and application for free surface flow in porous media. Int. J. Adv. Water Resour. **6**(1), 27–35 (1983)
21. Desai, C.S., Toth, J.: Disturbed state constitutive modeling based on stress-strain and nondestructive behavior. Int. J. Solids Struct. **33**(1), 1619–1650 (1996)
22. Desai, C.S., Basaran, C., Dishongh, T., Prince, J.: Thermomechanical analysis in electronic packaging with unified constitutive model for materials and joints, components, packaging and manuf. Tech. Part B, Adv. Packag. IEEE Trans. **21**(1), 87–97 (1998)

23. Desai, C.S., Dishongh, T., Deneke, P.: Disturbed state constitutive model for thermomechanical behavior of dislocated silicon with impurities. J. Appl. Phys. **84**(11), 5977–5984 (1998)
24. Desai, C.S.: Evaluation of liquefaction using disturbance and energy approaches. J. Geotech. Geoenviron. Eng. ASCE, **126**(7), 618–631 (2000)
25. Desai, C.S., Pradhan, S.K., Cohen, D.: Cyclic testing and constitutive modeling of saturated sand-concrete interfaces using the disturbed state concept. Int. J. Geomech. **5**(4), 286–294 (2005)
26. Sane, S., Desai, C.S., Jenson, J.W., Contractor, D.N., Carlson, A.E., Clark, P.U.: Disturbed state constitutive modeling of two Pleistocene tills. Q. Sci. Rev. **27**(3–4), 267–283 (2008)
27. Bak, P., Chen, K.: Self-organized criticality. Sci. Am. (1991)
28. Desai, C.S., Zaman, M.M., Lightner, J.G., Siriwardane, H.J.: Thin-layer element for interfaces and joints. Int. J. Num. Analyt. Meth. Geomech. **14**(1), 1–18 (1986)
29. Sharma, K.G., Desai, C.S.: An analysis and implementation of thin-layer element for interfaces and joints. J. Eng. Mech. ASCE, **118**, 545–569 (1992)
30. Desai, C.S., Ma, Y.: Modelling of joints and interfaces using the disturbed state concept. Int. J. Num. Analyt. Meth. Geomech. **16**(9), 623–653 (1992)
31. Fakharian, K., Evgin, E.: Elasto-plastic modeling of stress-path dependent behavior of interfaces. Int. J. Num. Analyt. Meth. Geomech. **24**(2), 183–199 (2000)
32. Desai, C.S., Basaran, C., Zhang, W.: Numerical algorithms and mesh dependence in the disturbed state concept. Int. J. Num. Methods in Eng. **40**(16), 3059–3080 (1997)
33. Desai, C.S., Zhang, W.: Computational aspects of disturbed state constitutive models. Int. J. Comp. Meth. Appl. Mech. Eng. **151**, 361–376 (1998)
34. A.K. Garakani, S.M. Hari, Desai, C.S., S.M. Ghafouri, B. Sadollahzadeh, H.H. Senejani. Testing and constitutive modeling of lime-stabilized collapsible loess, Pat I and II (Modeling and Validations). Int. J. Geomech. (Resubmitted After Review) (2018)
35. Xiao, Y., Desai, C.S.: Constitutive modeling for overconsolidated clays based on disturbed state concept: Part I and Part II. Int. J. Geomech. (ASCE, Resubmitted After Review) (2018)
36. Essa, M.J.K., Desai, C.S.: Cyclic testing and modeling of saturated interfaces with effect of surface roughness: I and II – Modeling and validations. submitt. Int. J. Geomech. ASCE (2018)
37. Toufigh, V., Shirkhorshidi, S.M., Hosseinali, M.: Experimental Investigation and Constitutive Modeling of Polymer Concrete and Sand Interface. Int. J. of Geomech. ASCE **17**(1) (2017)
38. Toufigh, V., Abyaneh, M.J.: Evaluation of rocks by considering softening behavior and volumetric deformation. Int. J. Geomech. (ASCE, Accepted for Publication) 2018
39. Essa, M.J.K., Desai, C.S.: Dynamic soil-pile interaction using the DSC constitutive model. Indian Geotech. J. **47**(2), 137–149 (2017)
40. Park, I.J., Desai, C.S.: Cyclic behavior of sand using disturbed state concept. J. of Geotech. Geoenviron. Eng. ASCE **126**(9), 834–846 (2000)

How to Improve Exchanges Between Academic Knowledge and Daily Practice?

Yasser El-Mossallamy[1(✉)], Gerhard Schulz[2], and Otto Heeres[2]

[1] Geotechnical Engineering, Ain Shams University, Cairo, Egypt
yasser.elmossallamy@arcadis.com
[2] Arcadis, London, UK

http://www.arcadis.com

Abstract. The rapid development in the last decades all over the world leads to the need of complex infrastructure projects as well as more and more high-rise buildings in complicated geotechnical conditions. Therefore, applications of highly sophisticated site investigation, design procedures applying complex numerical analyses and monitoring programs are now daily work of geotechnical engineers. New engineering generation are asked applying sophisticated numerical analyses to solve all geotechnical problems facing them during the design and construction of complicated structures. The interaction between the required enhanced theoretical knowledge with the appropriate methodologies for engineering applications is a big challenge for the young engineers.

The objective of this paper is to show the interaction between Academic Knowledge and Daily Practice regarding complicated and enhanced numerical analyses and real applications to reach the most economic design with minimum risks and also to optimize the required geotechnical solutions for some geotechnical applications.

Design of deep pit excavation as well as tunnel projects in complex geological conditions, huge sliding problem that takes place during construction of infrastructure projects and optimization of deep foundation of high rise buildings will be presented and illustrated as some practical examples of how sophisticated academic knowledge can be applied in complex megaprojects.

Keywords: Deep pit excavation · Tunnels · Huge sliding · High rise building Complicated geotechnical conditions · Monitoring · Numerical analyses

1 Introduction

A good understanding of the principals of geotechnical engineering with an adequate site investigation program is the basis of successful design and construction. In the history of civil engineers, man has built fantastic structures (Giza pyramids, Coliseum, Hagia Sophia, Taj Mahal etc.) with good engineering judgment. The new development in numerical analyses should help to optimize the projects and minimize the risks. Some examples concerning piled raft foundation, slope stability, deep pit excavation and huge earth work projects will be presented showing the interaction between

H. Shehata and C. S. Desai (Eds.): GeoMEast 2018, SUCI, pp. 32–47, 2019.
https://doi.org/10.1007/978-3-030-01926-6_2

engineering experiences, numerical analyses as well as monitoring programs to achieve the safety and economic aspects of structures.

1. **Piled raft foundation**

The piled raft foundation is a combined geotechnical system where the structure loads are resisted by both piles as well as the contact stress between raft and soil. The complex system consisting of raft, piles and soil should have sufficient bearing capacity (ultimate limit state). Nevertheless, the piles can be loaded near their limit geotechnical capacity. The serviceability requirements regarding the total and differential settlements as well as the tilting are the basic design criteria.

The history of application of piled raft started in Frankfurt, Germany in the 80 s of last century. The first building with piled raft foundation (Tor-Haus) was designed and constructed with very limited numerical analyses due to the limited availability of adequate soft-wares and computer capacity at that time. Nevertheless, the design was based on a long and enhanced experience with Frankfurt subsoil conditions via series of intensive researches in Darmstadt University. Also, a very enhanced and adequate monitoring system was applied to measure the performance of the foundation (settlement, load taken by piles, skin friction distribution along piles as well as raft contact stresses). The measurements have shown that 80% of the loads are taken by piles, piles in the middle of the building have very small load sharing and there is very high load concentration on the edge and corner piles (Fig. 1).

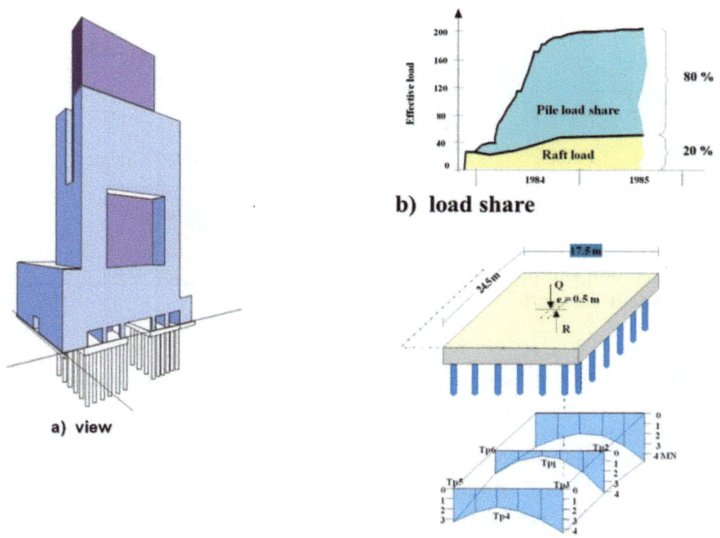

Fig. 1. The First piled raft foundation in Frankfurt, Germany (The Torhaus)

During the design of Messeturm, one has applied the experience gained from the Torhaus to reach an optimum design regarding the load share between piles and raft

(55% by piles) as well as an optimum mobilization of loads on all piles (more or less equal load distribution among all piles). Based on well engineering judgment, limited number of piles are applied in a three ring with the short piles at the outer ring and long piles in the middle ring to resist the Boussinesq's distribution that was observed by Torhaus. The general performance of the skin friction distribution along pile shaft agrees well with the measurements of the Torhaus and show a small development of the skin friction at the top part just beneath the raft. The skin friction increases with depth. The measured maximum skin friction in Frankfurt overconsolidated clay reaches about 160 kPa that is about 2.0 to 2.5 times the expected ultimate skin friction according to pile load tests on bored piles in similar overconsolidated clay. This phenomenon is due to the mutual interaction between raft and piles (Fig. 2). This is also a main feature of the performance of piled raft foundation.

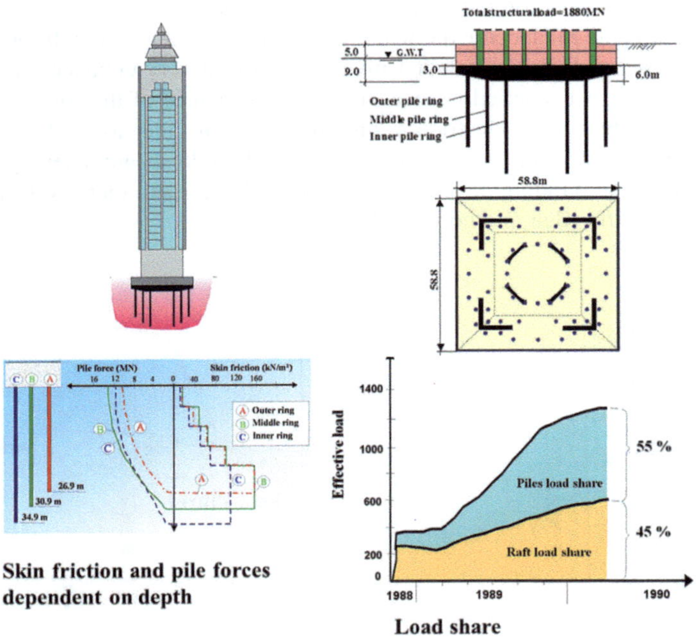

Fig. 2. The Messeturm high rise building in Frankfurt, Germany

The High rise buildings Messeturm and ML58 were the first high rise building where enhanced numerical models (El-Mossallamy 2009) were applied to back-calculate the foundation performance and start a new history of the application of numerical analyses to design piled raft foundations.

The detailed analyses have shown that the pile capacity of a single pile is completely different from the pile capacity of a pile as a member of a pile group and is still smaller than its capacity as a member of a piled raft foundation (Fig. 3). Limited number of piles (40 piles) can control the settlement to about 50% of its value as

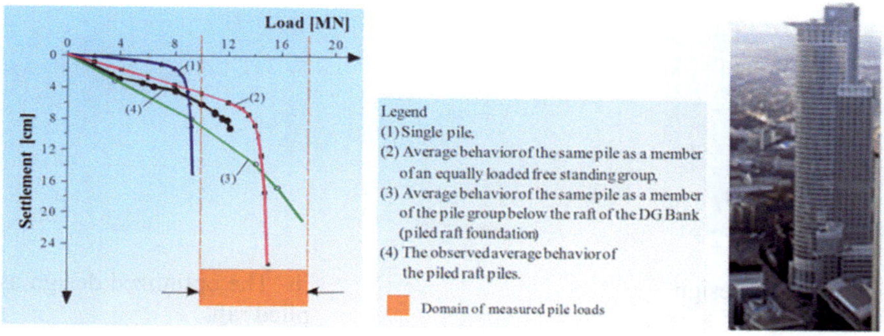

Fig. 3. Pile capacity as single pile, average of pile group and average of piles of piled raft.

conventional raft foundation. With only 40 piles, the piles can be located in an intelligent distribution just beneath the structure main loading elements (cores, shear walls and columns) to optimize the raft design compared with conventional raft or conventional raft on piles (Fig. 4).

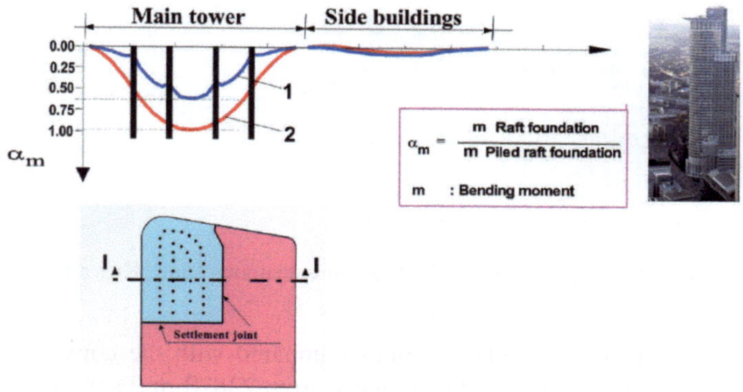

Fig. 4. The advantage of piled raft to optimize the straining action of the raft.

After having the unique experience of innovative applications of piled raft in Frankfurt, the piled raft foundation is applied in very complex geotechnical soil conditions (e.g. in Calcareous sands in Kuwait, Fig. 5).

The calcareous sand has low deformation parameters compared with the slice sand. The skin friction and end bearing resistance of large diameter bored piles in such soils will govern the design of the piled raft. A detailed pile load test on single pile under compression and tension were carried out to define the pile-soil parameters and to adjust the numerical model. The updated pile-soil parameters are then used to design the whole system in a real three dimension model (Fig. 6).

This design procedure leads to optimization not only of the pile depth to avoid penetrating in a deep aquifer and cause negative environmental impact but also to

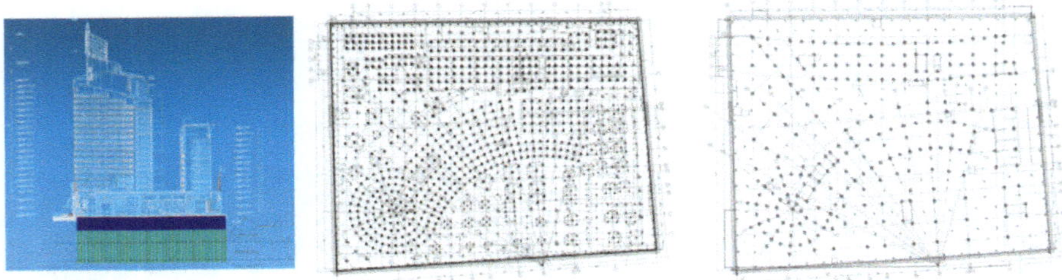

a- The basic design

b- The optimized design as piled raft

Fig. 5. Application of piled raft in calcareous sand, Kuwait.

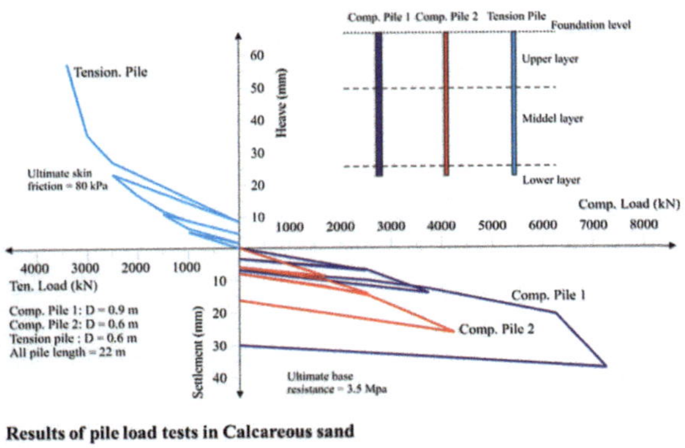

Fig. 6. Application of piled raft in calcareous sand, Kuwait.

significant reduction of the number of piles compared with the conventional raft on piles. The total length of piles could be reduced from 23870 m (basic solution) to only 6732 m (optimized piled raft).

Another area of application of piled raft was practiced in the design of the about 400 m high minarets in Haram in Makah, where the raft is partially supported on sound granite and partially on alluvium sediments. This is an innovative application of piled raft as a pure combination between conventional deep foundation and conventional raft foundation in one system (Fig. 7).

Again pile load tests are done to define the pile-rock parameters. A three dimensional model was developed to study the complex soil structure interaction. The pile stiffness under vertical and horizontal loads as well as the raft subgrade reaction module were determined and provided to the structural engineer for the design of the superstructure (Fig. 8).

Meanwhile the piled raft concept is also applied to optimize bridge foundations. For the Reichenbachtal-bridge in Germany, the subsoil conditions consisted of highly weathered rock to depth of about 50 m followed by slightly weathered rock formation.

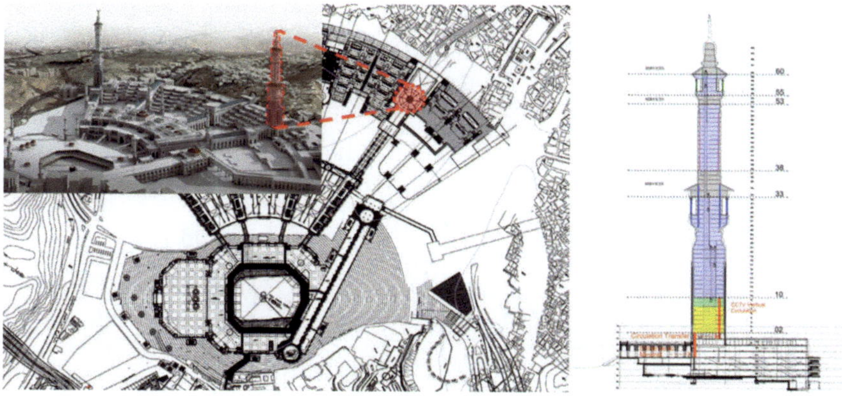

Fig. 7. Application of piled raft in Makah, Saudi Arabia.

In the basic design, long piles (longer than 50 m) were applied to penetrate the highly weathered rock formation and embedded in the slightly wreathed rocks. The application of the piled raft concept has led to dramatically optimization of pile lengths to only about 15.0 m. Again, detailed pile load tests are done to define the pile performance in the highly weathered rock. The results of the pile load tests are implemented in the numerical model for the design of the piled raft. The pile load tests need to demonstrate the entire load-settlement curve until failure. The measurements have proved the successes of the application (Fig. 9).

2. **Slope stability applying dowels**

The increase need of infrastructure leads to the construction of thousands km of highways and railways in different topographical and geotechnical conditions. Deep cuts are needed to achieve the required alignments. In many cases the cuts are done in variable geological conditions. According to site investigations, general design of the required slope inclinations and drainage systems are designed and applied. Nevertheless, a rest risk of unforeseen sliding planes can lead to some local instabilities due to existing sliding surfaces and/or weak zones. In such cases additional measures should be applied to achieve the required safety. Dowels of large diameter bored piles, diaphragm barrettes and shafts are used successfully to stabilize the slopes. Generally, the dowels are designed as cantilever rigid elements that sustain the driving forces of the slopes above the sliding surface and transfer these forces to the stable sub-ground beneath the sliding surface. This approach ignores the soil-structure interaction and leads to very expensive solutions. A very effective and efficient design methodology was developed considering the visco-performance of the soil. The studies done at the University of Karlsruhe (Germany) by Prof. Gudehus have shown that there is a relationship between the rate of slope movement and the factor of safety. By reducing the rate of movement to sufficient values, the factor of safety is increased ("performance based design") (Fig. 10).

Moreover, considering the deformation of the slope the dowels leads to definition of following conditions:

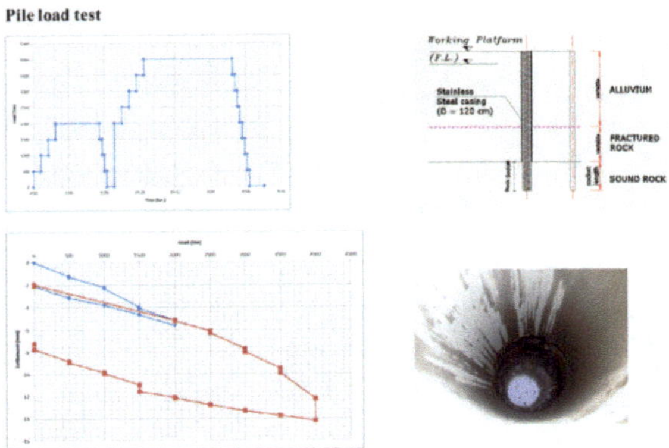

a- Pile load test on large diameter socketed bore piles

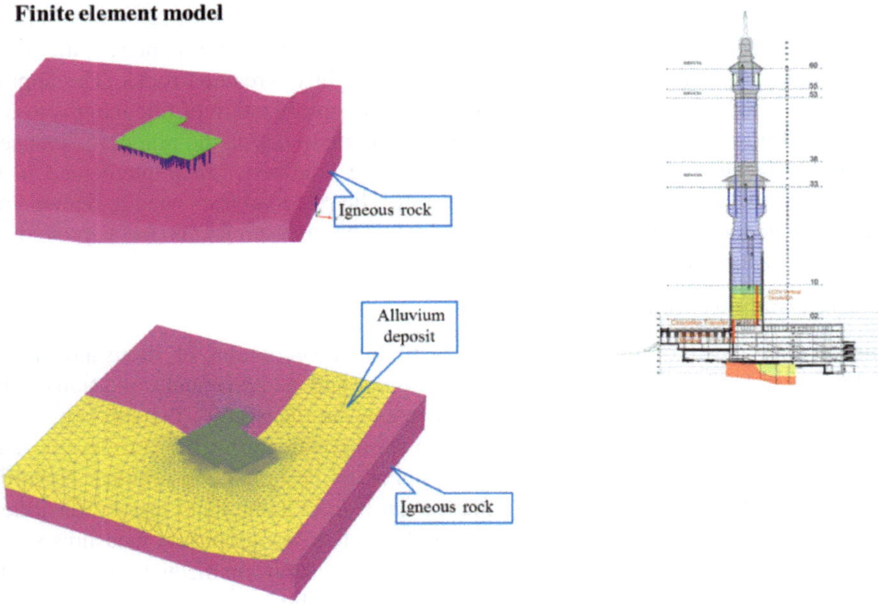

a- Three dimensional model

Fig. 8. Application of piled raft in Granite, Makah, Saudi Arabia.

- If the soil deformation is larger than that of the pile above the sliding surface, a conventional cantilever conditions will govern the design.
- If the soil movement is partially less than that of the piles, a neural plane can be defined at which the pile and soil movements are the same. Above this neutral plane, the soil movement will be less than that of the pile and hence the soil supports the piles. Beneath the neutral plane the pile movement is less than that of

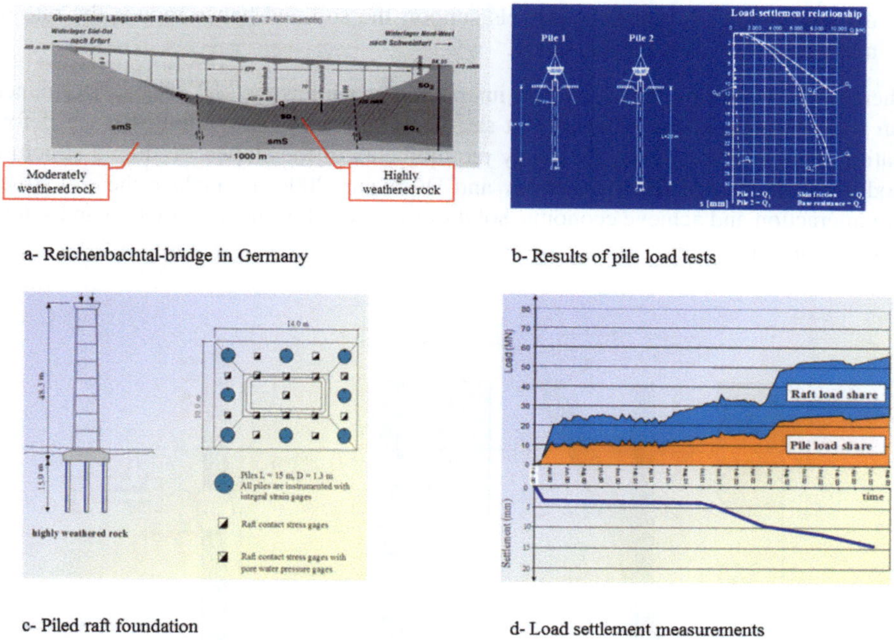

Fig. 9. Application of piled raft for bridge foundation in Germany.

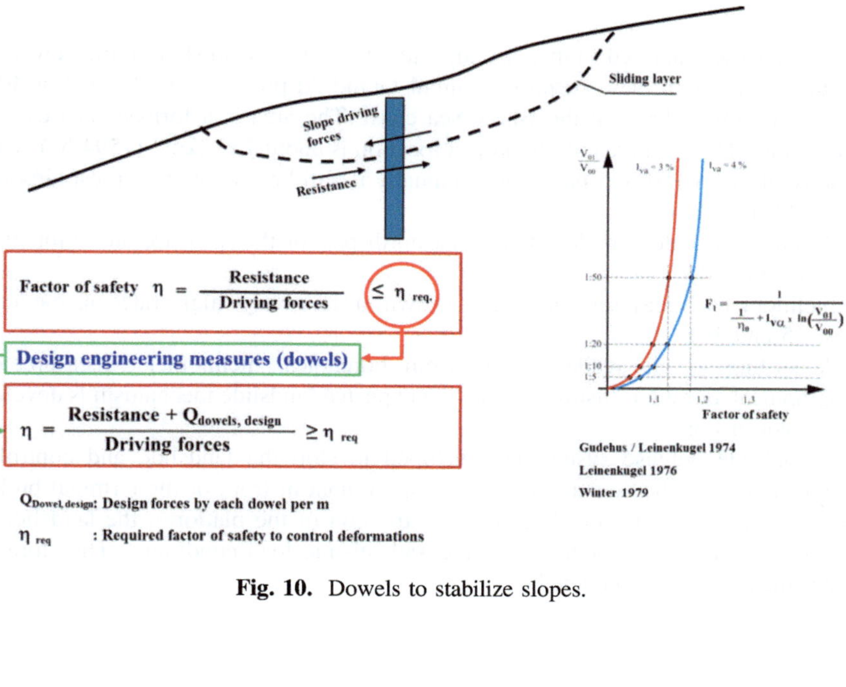

Fig. 10. Dowels to stabilize slopes.

the soil and in this zone the dowels support the soil and hence reduce the rate of movement and increase the safety.

Therefore, considering the soil-dowel interaction as well as the soil visco-performance can lead to an economic solution that satisfies the required limitation of slope deformation with time as well as the safety requirements of the slopes. A simple structure model was developed (El-Mossallamy and Dürrwang 2006) to analyze the soil structure interaction and achieve economic solutions that satisfies the deformation and safety requirements (Fig. 11).

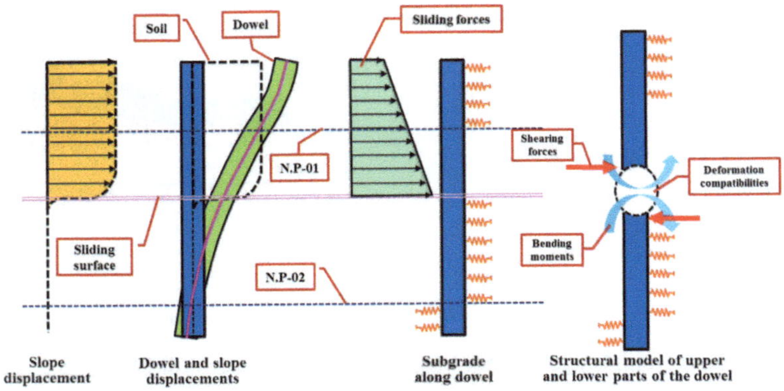

Fig. 11. Dowels theory.

This idea was applied to increase the stability of huge land slide in the new airport in Istanbul. Istanbul New Airport (Istanbul Grand Airport) will be located at 40 km north of Istanbul, direct at the Black Sea coast. The area is a former open-cast coal mining area. The total area of the project location is about 75.94km2 (7594 ha). With a capacity of up to 150 mln passengers annually it will be one of the largest airports in the world (Fig. 12).

A huge landslide was detected at the north part of the intermediate terminal pier (Figs. 13 and 14)

Inclinometer measurements have shown a relatively high rate of movement (Figs. 15 and 16).

According to all available data from boreholes, inclinometer measurements, topographical as well as historical data, the expected landslide mechanism is developed and shown in Fig. 17.

A multiple defense system is developed to stop the landslide and control the movement so that the construction of the apron area in front of the terminal building can be completed. After finishing the construction of the platform, the landslide stability will be achieved under both static and seismic load conditions. Therefore, following measures are conducted:

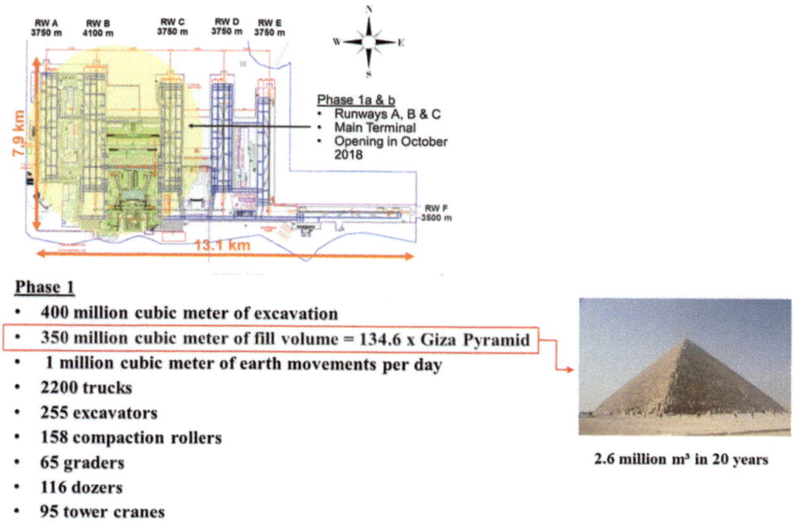

Fig. 12. Istanbul Grand Airport.

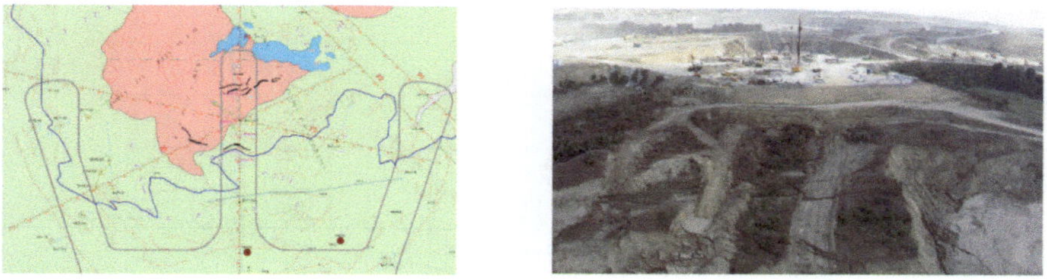

Fig. 13. Landslide at north pier.

Fig. 14. Sliding surfaces.

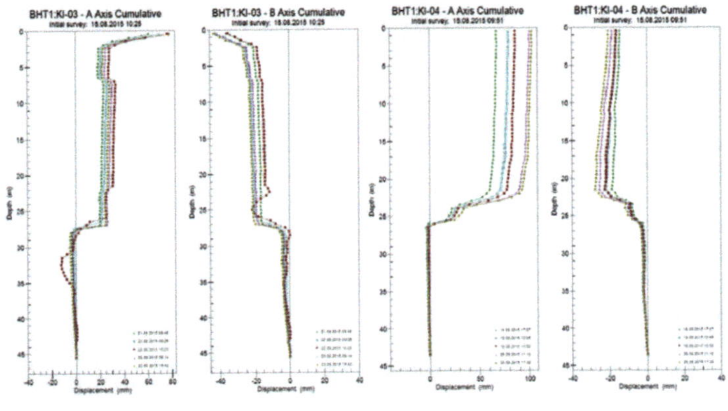

Fig. 15. Inclinometer measurements

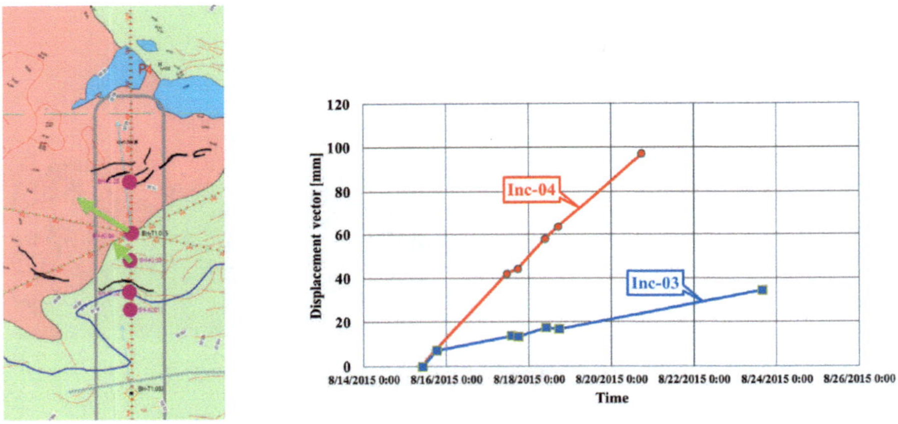

Fig. 16. Rate of movement

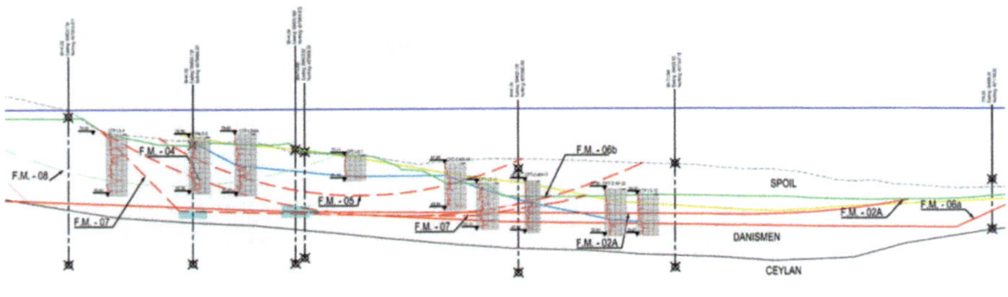

Fig. 17. Possible failure mechanisms

- Fill about 7.0 m in front of the landslide to stabilize the situation
- Construct the double pile rows cofferdam (Fig. 18)
- Remove all weak soils and sliding mass and stabilize the weak zone in front of the cofferdam via piles
- Construct the base drainage system
- Complete the embankment construction using cemented stabilized soil.
- The north part of the intermediate pier can be founded on shallow foundation rested on the cemented stabilized embankments and completely separated from the rest of the terminal building via settlement joint.

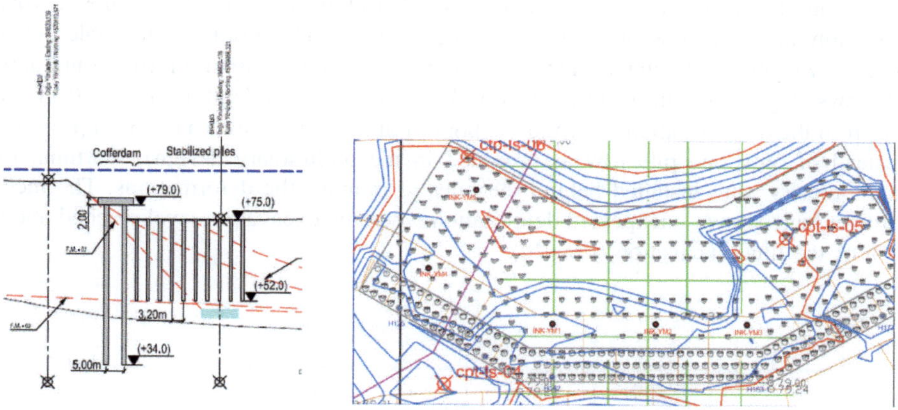

Fig. 18. Cofferdam solution

Figure 19 presets inclinometer measurements after the construction of the cofferdam that verifies the required stability during the construction of the apron platform.

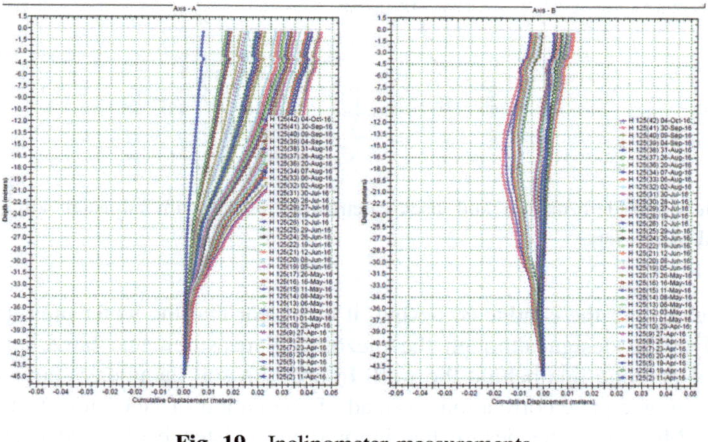

Fig. 19. Inclinometer measurements

3. Deep pit excavation

The need of underground subways is increased in the last decades especially in Mega cities. The very limited available time for design and construction with the need of economical solution can affect the project progress in a negative way. In such cases a good engineering judgment with adequate monitoring system can minimize risks and save time and money. The construction of a new metro line requires the construction of underground metro stations in very complex subsoil conditions and sensitive neighborhoods. The middle station is the main point for the construction of the whole metro line as four TBM machines should start the tunneling progress from this station in the two opposite direction. The depth of the station reaches about 30 m; the upper soil layer consists of clay and silty clay/clayey silt followed by highly weathered rock formation and slightly weathered rock in bigger depth. The ground water table lies in the clayey layer. In the final design – as a supposed economic solution- two contiguous pile rows staggered with depth (one from the surface up to a depth of about 20 m and one from the depth of about 20 m to the bottom of the excavation) have been provided (instead of only one pile row as provided in the preliminary design).. Fortunately, inclinometers were installed behind the piles to monitor the deformations. The inclinometer have shown unexpected deformation when the excavation level reached about 60% of the final excavation (Fig. 20).

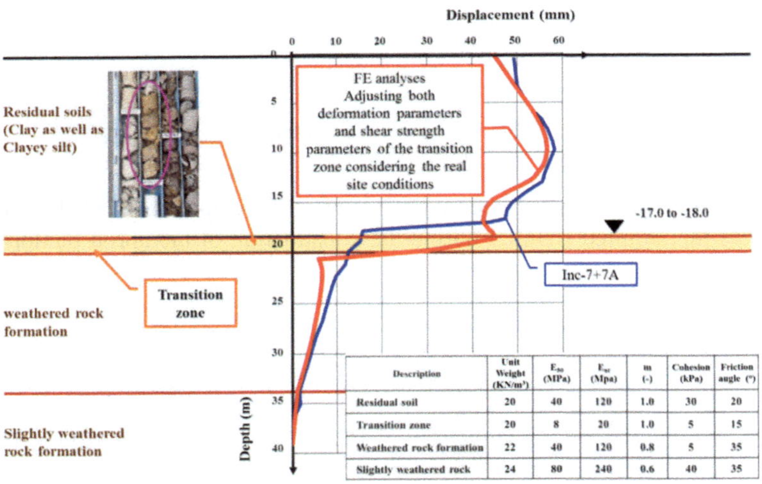

Description	Unit Weight (KN/m³)	E_{50} (MPa)	E_{ur} (Mpa)	m (-)	Cohesion (kPa)	Friction angle (°)
Residual soil	20	40	120	1.0	30	20
Transition zone	20	8	20	1.0	5	15
Weathered rock formation	22	40	120	0.8	5	35
Slightly weathered rock	24	80	240	0.6	40	35

Fig. 20. Measured deformation and back-calculation results with adjusted soil profile and geotechnical parameters.

After reviewing the conducted design, it was clear that the water pressure was not considered in the design although the piezometer measurements have approved the existence of phreatic line behind the wall. In addition soil parameters have been used by applying in adequate correlations instead of evaluation of laboratory tests and local experience. Moreover, the design has ignored the existence of a weak layer at the transition zone between the upper silty clay/clayey silt and the highly weathered rock formation. Considering the groundwater and the weak layer, it was clear that the basic design cannot sustain the forces induced during final excavation. Additional measures mainly longer anchors with higher prestressing forces and internal struts are added to limit the deformations and achieve the required safety.

Another problem was the construction of a shaft on another station, after fishing the 15.0m diameter shaft and starting the tunnel excavation in NATM, large displacement are observed (Figs. 21, 22 and 23).

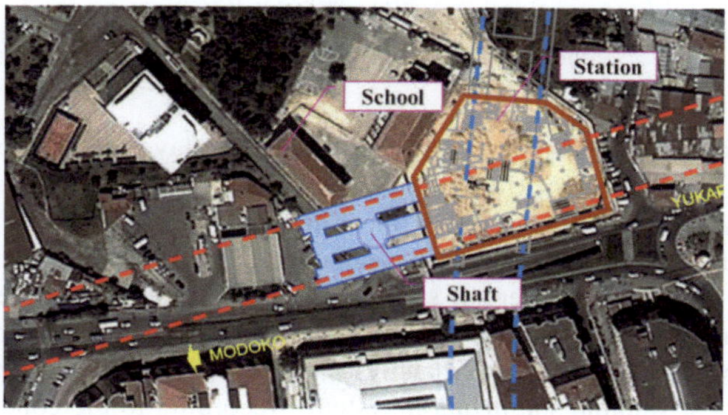

Fig. 21. Dudullu-station, Shaft N1, Istanbul, Turkey

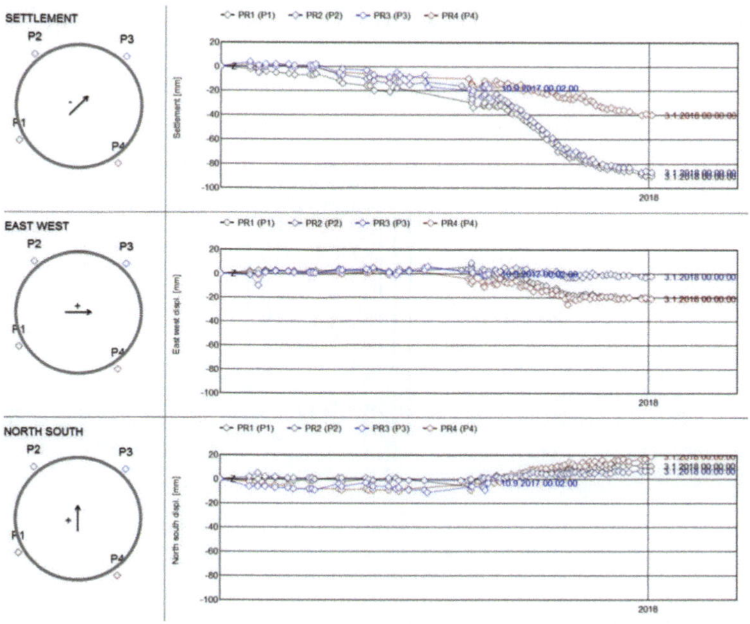

Fig. 22. Shaft displacements

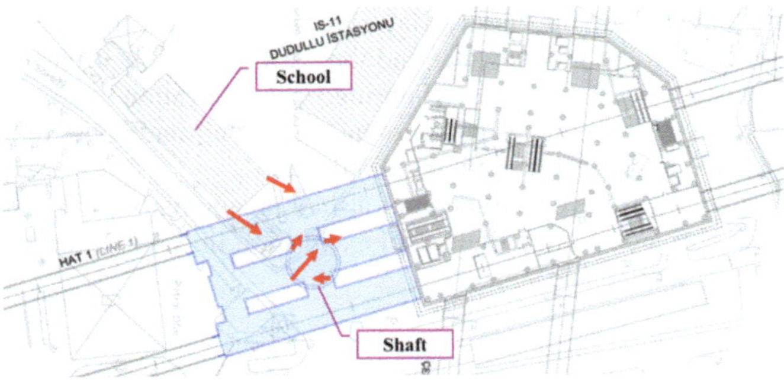

Fig. 23. Induced displacement of the neighboring school

The sub-ground conditions consist of a thick upper layer of stiff to hard overconsolidated clay followed by rockformation of medium to strong limestone with different degrees of weathering. The surface of the limestone is very wavy and its elevation varies rapidly in very short distances. A weak clayey layer was detected at the interface clay/limestone and consist potential sliding surfaces depending on the dipping angle and dipping directions related to excavation and the induced stress changes.

Three dimensional finite element analyses were carried out to model the shaft and to design the most suitable stabilization system. According to the conducted analyses, nine large diameter bored piles are designed as four groups to support a capping raft connected with the capping beam of the shaft to transfer the shaft loads to either the limestone at three corners or as floating piles at the fourth corner where no limestone can be detected down to 55.0 m below existing ground surface (Fig. 24).

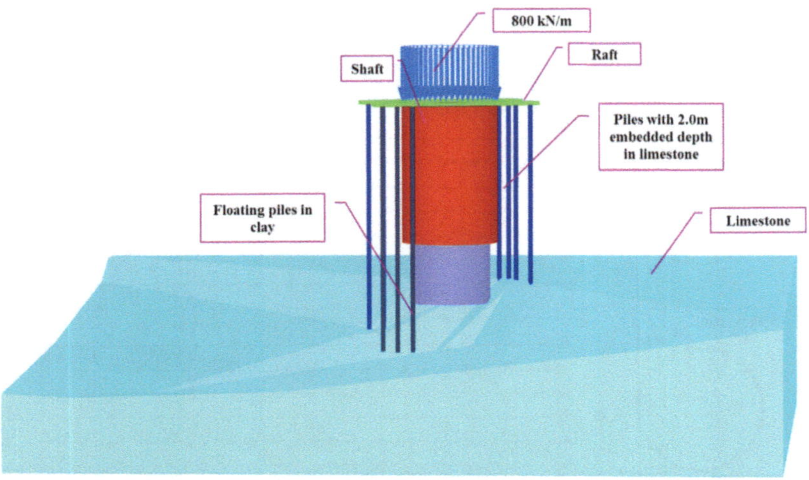

Fig. 24. Modeling the shaft, the stabilization piles and the raft

2 Conclusions

Computers with high capacity and advanced software with sophisticated constitutive laws are nowadays accessible for almost every engineer. This can enable more economic safe design and construction of structures than decades before. But only, if these innovative calculations go hand in hand with geotechnical understanding and engineering judgment. The calculations need realistic input parameters based on adequate and accurate site investigation.

It is the responsibility of the geotechnical consultant to insist in sufficient geotechnical investigations (quality instead of quantity), to demonstrate the risks to the stakeholders, to minimize risks and to develop economic solutions. This is not conflict.

The mutual cooperation between theory and applications with good engineering understanding and judgments can help to optimize the geotechnical solutions and minimize the risks. Enhanced and complicated constitutive laws and numerical analyses can be very useful but cannot stand alone and replace the experience and the basic understanding of soil mechanics.

Monitoring the real performance of structures and back-calculation are essential to prove the assumptions and to enhance the engineering understanding to be applied in further projects. Large in-situ tests can be very helpful to determine the real performance of the soil and hence improve the design and construction to reach the most economic solution.

Acknowledgments. The first author would like to emphasize his thanks to the geotechnical institute at TU Darmstad where he has done his PhD work under the supervision of Professor Eberhard Franke and Professor Rolf Katzenbach. Special thanks should be awarded to Professor Katzenbach for the many years cooperation in many national and international projects.

The authors appreciate the support of all our colleagues in Arcadis during the last 25 years specially Mr. Duerrwang and Mr. wittman from Arcadis germany.

References

El-Mossallamy, Y.M., Lutz, B., Duerrwang, R.: Special aspects related to the behavior of piled raft foundation. In: International Conference on Soil mechanics and Geotechnical Engineering 17th ICSMGE, Alexandria (2009)

El-Mossallamy, Y.: Employing "Piled Rafts in Gulf Countries" Piling and Deep Foundations, Summit 2009, Dubai, UAE, 1st and 2nd March 2009

El-Mossallamy, Y., Dürrwang, R.: Bauwerke in instabilen Hängen – Fallbeispiele" 29. Baugrundtagung, 27–30 Sept 2006 in Bremen

Numerical Investigation of the Frequency Influence on Soil Characteristics During Vibratory Driving of Tubular Piles

Reza Daryaei(✉), Montaser Bakroon, Daniel Aubram, and Frank Rackwitz

Chair of Soil Mechanics and Geotechnical Engineering, Technische Universität Berlin, Berlin, Germany
r.daryaei@campus.tu-berlin.de

Abstract. In offshore geotechnics, tubular piles are commonly used as the foundation system. Such piles are installed using vibratory or impact driving. The choice of the proper loading configuration plays an important role in the driving performance, especially in reaching the desired penetration depth. Numerical evaluation of such processes involves handling large material deformation, making it hard for the classical numerical methods to reach a reliable result after significant deformation. In addition, in case of the dynamic cyclic loading, the soil exhibits complex behavior which emphasizes the role of a suitable soil constitutive equation. In this study, a numerical model is developed and utilized to evaluate the effects of the frequency in the vibratory installation of tubular piles on the neighboring soil. The numerical model employs the robust Multi-Material Arbitrary Lagrangian-Eulerian (MMALE) method in conjunction with an advanced material model formulation based on the hypoplasticity concept, and is validated against an experiment done at TU Berlin. Subsequently, a parametric study is performed by applying six different frequencies between 12 and 30 Hz to the dynamic load. The resulting penetration depth, void ratio and the lateral stress distribution in the soil are compared and evaluated. It is concluded that an optimum frequency must be determined to reach the maximum penetration depth by using the same load magnitudes.

1 Introduction

Tubular piles are one of the most-practiced types of deep foundation systems, especially for offshore applications. The installation performance of such piles is influenced by many factors, such as pile diameter and length, soil strength, pile-soil interaction, and the driving method. Conventionally, impact or vibratory driving is applied. Compared to impact driving, the vibratory driving produces less noise level, induce less damage to the pile, and offers a significantly higher penetration rates (Holeyman and Whenham 2017; Rausche 2002).

In this study, the effects of vibratory driving on the soil are investigated by varying the installation frequency. The soil response is analyzed by evaluating the resulting void ratio and lateral stress distribution. Due to the complex soil behavior and the large

H. Shehata and C. S. Desai (Eds.): GeoMEast 2018, SUCI, pp. 48–61, 2019.
https://doi.org/10.1007/978-3-030-01926-6_3

soil deformations during installation, a suitable numerical formulation is required to avoid the inevitable inaccuracies due to mesh distortion. Therefore, the Multi-Material Arbitrary Lagrangian-Eulerian (MMALE) method is employed to address this issue (Aubram et al. 2017; Benson 1992). In addition, owing to the high dependency of the numerical model accuracy on the soil material model, a robust constitutive equation based on the hypoplasticity concept is adopted (von Wolffersdorff 1996). There are numerous examples of successful simulations of complex problems using the MMALE and the hypoplastic material model in the literature (M. Bakroon et al. 2018a).

Previous works regarding vibratory driven piles consist of both numerical and experimental studies. One of the early studies on penetration rate was done by Feng and Deschamps (2000), where they derived an empirical equation for a penetration rate of a pile which is installed using hammer driving. In this equation, various effective factors on pile driving were considered.

In a series of work done by Henke and Bienen (2013, 2014) and Henke and Grabe (2008), the effect of various installation methods such as quasi-static jacking, impact and vibratory driving on soil plugging, a phenomenon where the soil goes up inside the tubular pile and increases the penetration resistance, were both numerically and experimentally evaluated. They concluded that the driving method affects the soil behavior in terms of the formed soil plug inside the pile.

Heins et al. (2016) simulated the vibratory pile installation using the so-called zipper modeling technique, where a very small tube is defined inside the soil and on the pile axis which is initially in contact with soil. During the penetration, the soil looses its contact with the tube and starts to interact with the pile. Despite handling the induced large deformation due to pile penetration, this technique is reported to have some drawbacks regarding soil drifting and twisting (S. Henke 2010).

Different installation frequencies and driving energies were applied to the pile. Results consisted excess pore water pressure values as well as the resulting penetration depth. It was concluded that by applying more frequency and low energy, more penetration is achieved.

In a recent study done by Qin et al. (2017), numerous field tests were compiled to assess the effect of seven influencing parameters in the vibratory driving system on penetration rate of three different sheet pile sections. Results suggest that the effect of vibratory driving becomes significant at later stages of the penetration.

The structure of this paper is as follows. First, the theoretical background of the numerical model considerations including the chosen element formulation and the soil material model are presented. Then, the developed numerical model is described. The model is validated against an experiment followed by a parametric study on frequency effects on the surrounding soil. Subsequently, the obtained results from the parametric study are presented and discussed. Concluding remarks and outlook on future works are given at the end.

2 Theoretical Background

This section gives a brief description of the utilized material model, the numerical approach, and the contact formulation used to simulate the soil-structure interaction.

2.1 The Hypoplastic Material Model for Granular Materials

The hypoplastic material model developed by von Wolffersdorff (1996) and improved by Niemunis and Herle (1997) is used in this study. Generally, the idea behind the hypoplasticity concept is that it does not distinguish between elastic and plastic deformation (Kolymbas 1977). However, the first developed models exhibited shortcomings in capturing the realistic soil behavior under cyclic loading. The strain accumulated at each load cycle, resulting in a very significant strain accumulation compared to the experimental measurements. This issue was addressed by an intergranular strain tensor introduced as an additional state variable which represents the strain history (Niemunis and Herle 1997).

Unlike elastoplastic material models, the hypoplastic constitutive model is defined by a single incrementally nonlinear equation. The stress rate of the granular material, T, is determined by the effective stress, T, intergranular strain, δ, and the void ratio, e (Niemunis and Herle 1997):

$$\dot{\mathbf{T}} = \mathcal{M}(\mathbf{T}, \mathrm{e}, \delta) : \mathbf{D} \tag{1}$$

The void ratio in the Eq. (1) is constrained by e_i and e_d which reflect the minimum and maximum void ratio, respectively. Another limiting parameter is the void ratio at the critical state, e_c. To adapt the limiting void ratio parameters based on the current mean pressure a granular hardness parameter, h_s, is defined.

Hypoplasticity predicts the nonlinear and inelastic behavior of granular materials such as dilatancy, owing to its dependency on the void ratio of the soil. Due to its robustness observed in previous works, this material model is also used in the present study.

2.2 The MMALE Numerical Approach

Problems involving large soil deformation are often inevitable in geotechnical engineering. For instance, pile penetration, slope failures and avalanches, and liquefaction are major areas of interest during the last decades (Aubram et al. 2017). Numerical analysis of such problems is challenging, particularly due to the complex material behavior. The Finite Element Method (FEM) can be generally considered as a suitable numerical approach to address these problems. In FEM, there are two principal viewpoints to treat material deformations, namely the Lagrangian viewpoint and the Eulerian viewpoint.

In the Lagrangian viewpoint, the computational grid is pinned to the material particles, meaning that if the soil deforms, the grid deforms accordingly. In this approach considerable shortcomings arise when the soil significantly deforms, i.e. elements may encounter large distortion which leads to the solution divergence or unreliable results (Belytschko et al. 2000). On the other hand, in the Eulerian viewpoint, the grid is fixed and the material moves freely through the grid. Although large deformations and vorticity issues are addressed, the Eulerian approach requires extra considerations for treating path-dependent material behavior and tracking material interfaces (Benson 1992).

A new class of methods achieved to combine these two approaches to gather the advantages of both viewpoints while minimizing their shortcomings. Considering the grid-based methods, the most promising approaches include the Simplified or Single-Material Arbitrary Lagrangian-Eulerian (SALE), Coupled Eulerian-Lagrangian (CEL), and Multi-Material Arbitrary Lagrangian-Eulerian (MMALE) methods (Benson 1992).

In SALE, remeshing/rezoning after each Lagrangian step is performed. In a comprehensive work done by Aubram (2013, 2015) this method is thoroughly discussed and evaluated. Bakroon et al. (2017) studied the achieved performance in numerical calculations by using SALE, compared to the Lagrangian explicit and implicit methods The efficiency of SALE method in the aspect of both accuracy and mesh improvement was proved for problems associated with moderate deformation. However, for large deformation, this method still holds drawbacks. This drawback is caused by the material boundary/interfaces which can only slide along the material boundary/interface. For a more detailed discussion the reader is referred to (Aubram et al. 2017; M. Bakroon et al. 2018b).

Therefore, alternative methods such as CEL and MMALE have been developed which both employ a non-aligned grid with material boundaries and interfaces. Thus, some grid elements may contain a mixture of two or more materials which is referred to as multi-material elements. In both methods, a material-free or void zone must be defined within the grid holding neither mass nor strength such that the materials can flow into these regions of the physical space. In both CEL and MMALE, after performing one or several Lagrangian steps, the mesh is rezoned to its initial configuration to maintain mesh quality (rezoning/remeshing step). In case of MMALE, a new arbitrary mesh is developed which is different from the initial mesh configuration. Subsequently, the solution is transported from the deformed mesh to the updated or original mesh (remapping/advection step). Figure 1 illustrates the differences between the aforementioned methods. The sub-steps are not performed simultaneously but in series using the operator-splitting technique (Benson 1992).

A recent study conducted by Bakroon et al. (2018b) assessed the feasibility of CEL and MMALE methods in realistic geotechnical large deformation problems in comparison with SALE and classical Lagrangian methods. It was concluded that MMALE and CEL can be considered as promising candidates for solving complex large deforming problems. The applicability of MMALE in conjunction with a complex soil material model was investigated in another work done by Bakroon et al. (2018a).

2.3 Pile-Soil Interaction

To define the interaction between the pile and the soil, a robust contact scheme should be introduced. Usually, in CEL and MMALE formulation, the structural part and the soil are defined as the Lagrangian and Eulerian part, respectively. Penalty contact is generally employed due to its simplicity and robustness. The interaction between Lagrangian and Eulerian nodes is modeled as springs whose seeds and anchors are attached to the Lagrangian and Eulerian nodes, respectively (Benson and Okazawa 2004).

A schematic view of the penalty contact scheme is shown in Fig. 2 where two springs for both normal and tangential directions (k_t and k_n, respectively) are defined. The contact forces are calculated by considering a small amount of penetration, using

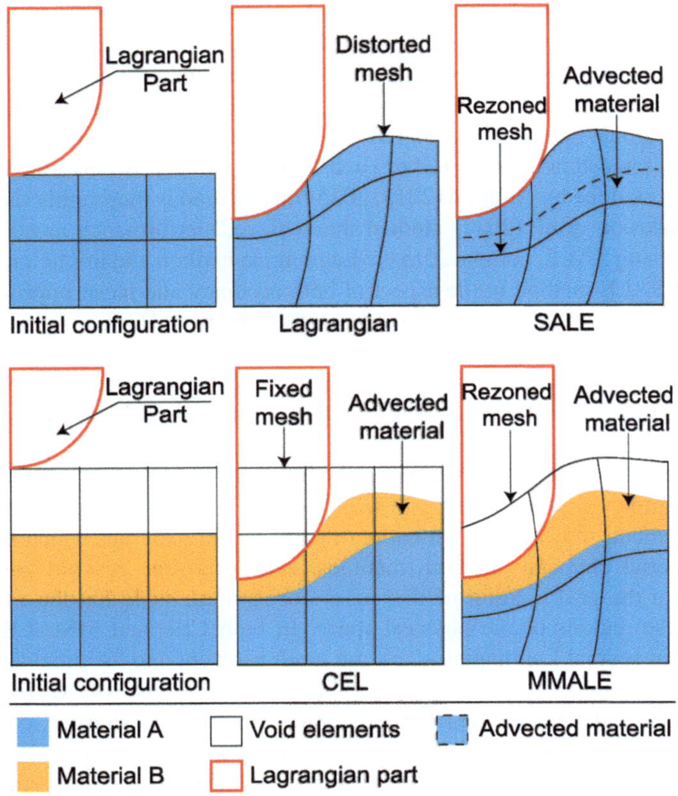

Fig. 1. Schematic diagram of CEL, SALE, and MMALE approaches comparing the effect of mesh rezoning (remeshing) and remapping (advection) steps of the solution.

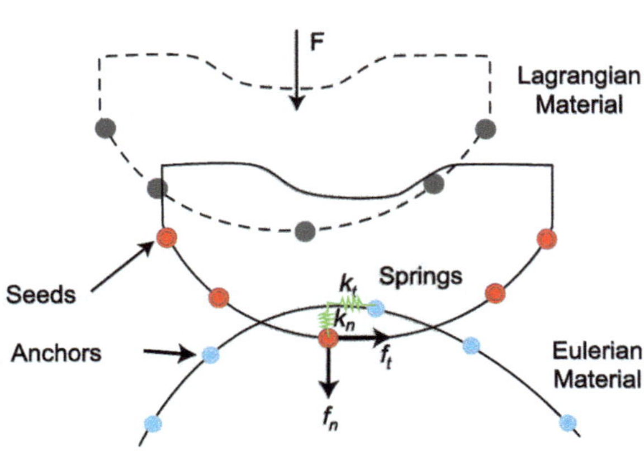

Fig. 2. Schematic view of penalty contact scheme

the equation $F_p = k_p d_p$, where d_p is the penetration depth, F_p is the spring force consisting of two principal forces, f_n, and f_t for normal and tangential direction, respectively (Benson and Okazawa 2004).

3 Numerical Model

3.1 Description of the Model

In this section, a description of modeling considerations using MMALE technique in LS-DYNA/Explicit is presented where a pile is installed in the soil using a vibratory force. An axisymmetric model is developed to reduce the computational costs. The model configuration is shown in Fig. 3a. The load history curve of the vibratory force is depicted in Fig. 3b.

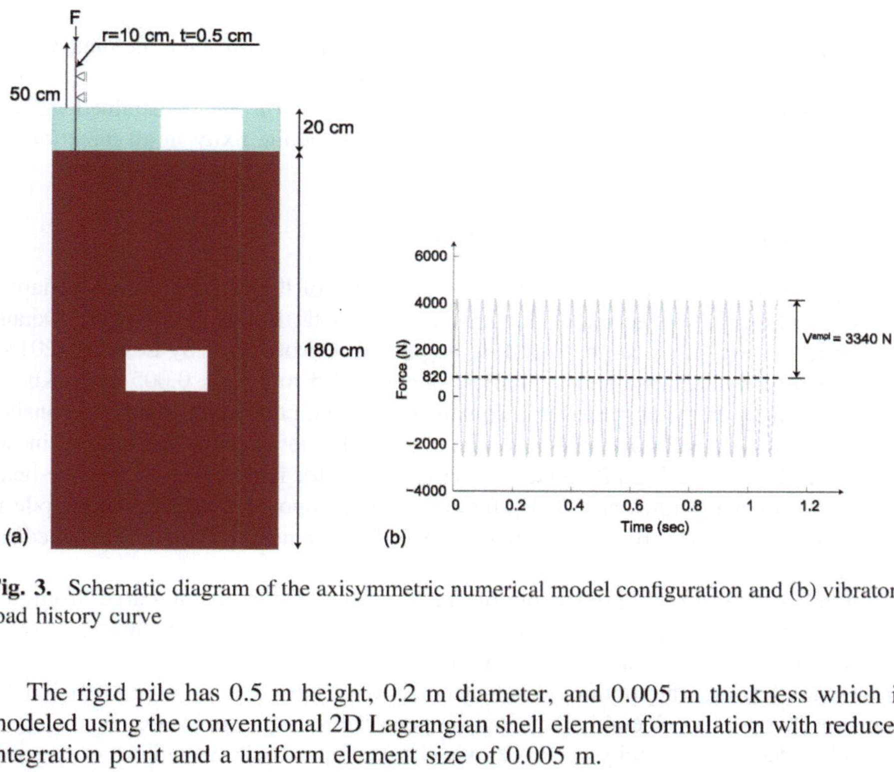

Fig. 3. Schematic diagram of the axisymmetric numerical model configuration and (b) vibratory load history curve

The rigid pile has 0.5 m height, 0.2 m diameter, and 0.005 m thickness which is modeled using the conventional 2D Lagrangian shell element formulation with reduced integration point and a uniform element size of 0.005 m.

For the soil, a mesh with 2 m height and 1 m radius with the one-point integration MMALE shell element formulation is generated. The equipotential smoothing technique is applied to the mesh nodes to maintain the mesh quality (Winslow 1963). For the advection step, the 2nd order accurate van Leer method is chosen (Van Leer 1977).

An unstructured mesh, ranging from 0.005–0.05 m element width is used. The mesh contains the soil up to the height of 1.8 m. A void domain with 0.2 m height, which has neither mass nor strength, is defined above the soil material to enable the soil to move to this domain after penetration starts. The hypoplastic material model is adopted, whose corresponding material constants for Berlin sand are listed in Table 1. The relative density of the soil is I_D = 75% ($e_{initial}$ = 0.465). The initial stress in the soil is defined with assigning the gravity acceleration as 10 m/s^2 and using the lateral earth pressure, K_0 = 0.5.

Table 1. Hypoplastic material constants for Berlin sand

φ_c [°]	h_s [MPa]	n	e_{d0}	e_{c0}	e_{i0}	α	β	m_R	m_T	R	χ	β_r
31.5	230*	0.3	0.391	0.688	0.791	0.13	1	4.4	2.2	1×10^{-4}	6.0	0.2

*The actual value of granular hardness, h_s, is 2300 MPa. This value is reduced by 10% due to low-stress soil state

To define the coupling between pile and soil, penalty contact is defined with a tangential friction coefficient of 0.2 which corresponds to tan(ϕ/3). The pile is fixed against horizontal movements. The lateral sides of the soil are constrained against movements in a direction perpendicular to their faces, while fixity in all directions is applied to the bottom of the soil.

3.2 Validation Against Experimental Results

An experimental test was conducted at the laboratory of the Chair of Soil Mechanics and Geotechnical Engineering at Technische Universität Berlin (TU Berlin). Details regarding the test set-up can be found in the published work done by Le et al. (2018). The experiment includes a half-cylindrical pile with 1.5 m length, 0.005 m thickness, and 0.2 m outer diameter placed in a container filled with the Berlin sand and consists of three rigid steel walls and one glass panel. The pile movement is constrained in the horizontal direction using pile guides. A vibratory motor is installed on the pile head with 1670 N driving force and 23 Hz frequency. The imposed dead load on the pile is about 410 N. To measure pile penetration, two displacement sensors are installed on the pile.

Figure 4 shows the comparison of the resulting displacement curve from the numerical model compared against experimental measurement. At early stages of driving, the penetration rate is significant due to the less soil resistance and confining pressure. Afterward, the frictional force increases since the resulting force normal to the pile skin increase considerably, causing the penetration rate to decrease.

In Fig. 5a, b, the velocity vectors at two-time stamps are shown which corresponds to the times when the pile pushes the soil and when the pile is pulled out (see Fig. 5c). During the push as shown in Fig. 5a, the soil particles under the pile tip tend to move to lateral sides and then to the upward direction. On the other hand, the particles near the pile shaft are moved with the pile due to the friction. The soil motion inside the pile (i.e. the left side of the pile in the model) is more significant than outside the pile. This can

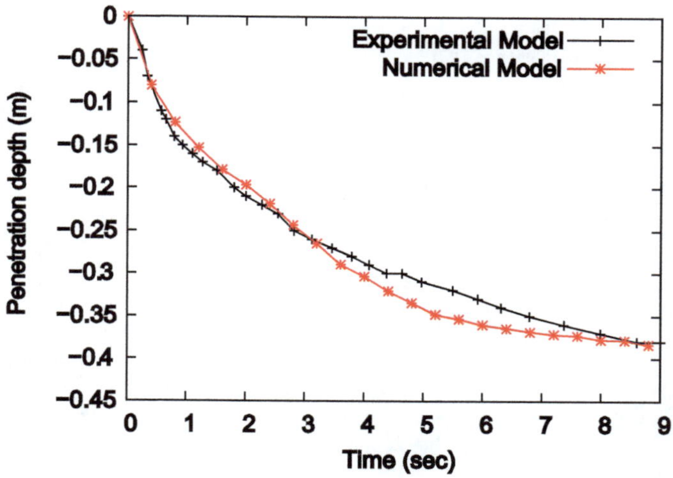

Fig. 4. Penetration depth vs. time curve obtained from numerical model and experimental measurement

be argued to be caused by the more confinement stress inside the pile. Moreover, the soil motion inside the pile is rather uniform, which seems like that the soil moves as a block according to the pile motion. This point is mentioned in several works in literature and is referred to as the "soil plugging", a phenomenon where the soil inside the pile moves as a block along with pile during the penetration (De Nicola and Randolph 1997). During the pile pull out stage as shown in Fig. 5b, the soil movement direction reverse as the pile tends to go up. Again, the soil inside the pile moves as a block with a more uniform velocity than the soil outside the pile, but with less magnitude compared to the pushing stage. The reversed movement is caused by releasing the pressure resulted by the pile push in the previous stage. Hence, the soil tends to return to its previous and more stable state which is similar to what is observed in loading-unloading tests. Concerning the points discussed above, it can be argued that the numerical model captures reasonably the soil condition in the experiment.

3.3 Parametric Study

To study the effect of vibration on the driving performance, six models with various frequencies are developed and compared. Both the static and dynamic load amplitudes are maintained for all models while the frequencies are changed to 12, 16, 19, 23, 27 and 30 Hz. All other considerations including model geometry and element size, initial state and boundary conditions are also maintained as those used in the previous section.

The penetration curves for all models are plotted against time in Fig. 6. The difference between the penetration trends becomes significant after about 4 s which corresponds to approximately 0.3 m. It may be argued that the soil at the lower depth did not show a high resistance against penetration. Therefore, this part of the penetration can be mainly attributed to the force induced by the static load.

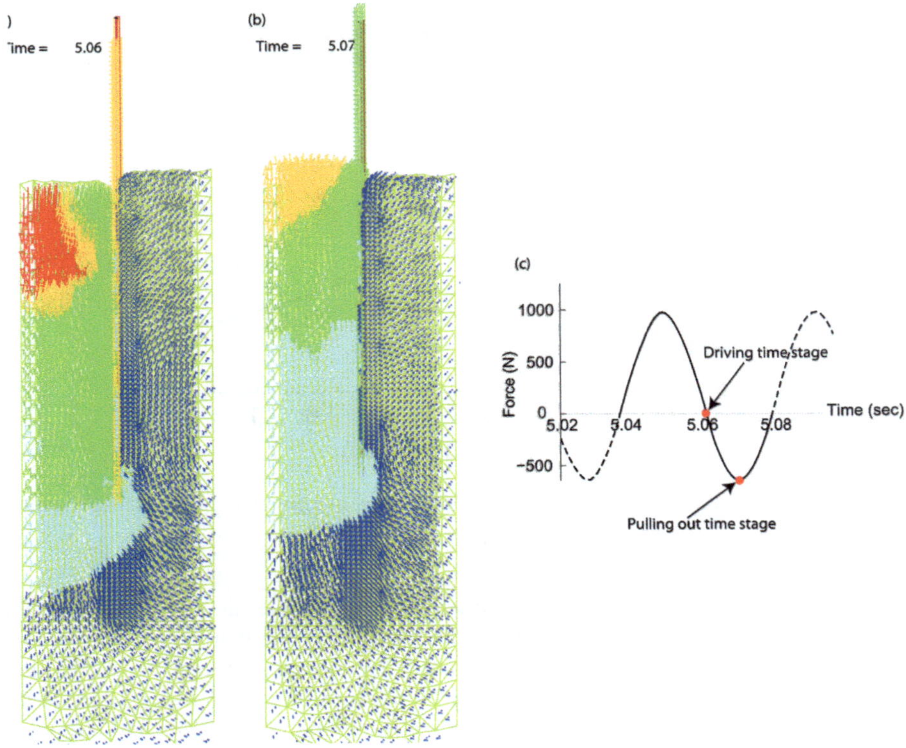

Fig. 5. Velocity vector of the soil regime during pile (a) driving, (b) pulling out and (c) the corresponding loading time stages for the validation model

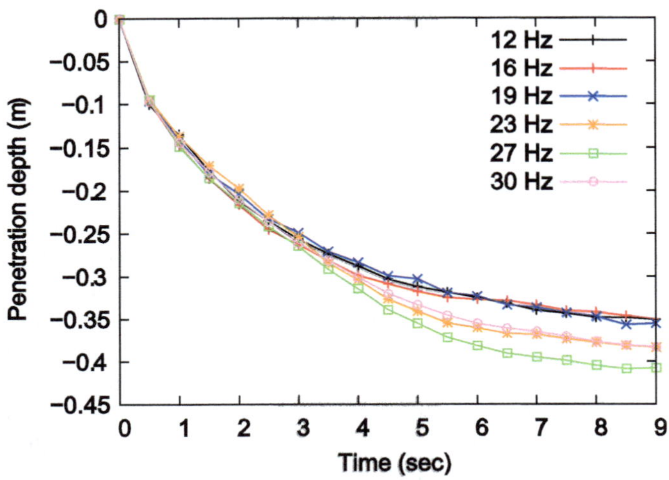

Fig. 6. Penetration curves of the models with different frequencies versus time

It is observed that the trend and the final value of the penetration is identical for lower frequencies of 12, 16, and 19 Hz. In addition, the models with 23 and 30 Hz resulted in a more penetration value than those with lower frequencies. Finally, the model with 27 Hz gave the most penetration depth.

During the vibratory installation, the soil is affected in several ways. First, the area around the pile is partly loosened due to the disturbance enforced by the pile vibration. This makes the pile penetration easier since the disturbed area exhibits lower resistance. Second, the soil regime located further from the pile is densified due to the same reason, making the soil grains to reach a more compacted arrangement. This might adversely affect the driving performance since the compacted soil shows more confinement and resistance, which leads to less penetration. Hence, an optimized frequency shall be assigned which results in more penetration without significant resistance.

To assess the compaction and loosening, the void ratio contours are shown in Fig. 7. It can be observed that by increasing the frequency, the compacted soil volume increases as well. On the contrary, the growth of the loosened soil regime is not as significant as the densified area. Thus, by using a higher frequency a larger area is compacted which resists more against the penetration and therefore reduces the disturbance effect which facilitated the penetration. Consequently, using the higher frequency may not always result in a more penetration depth.

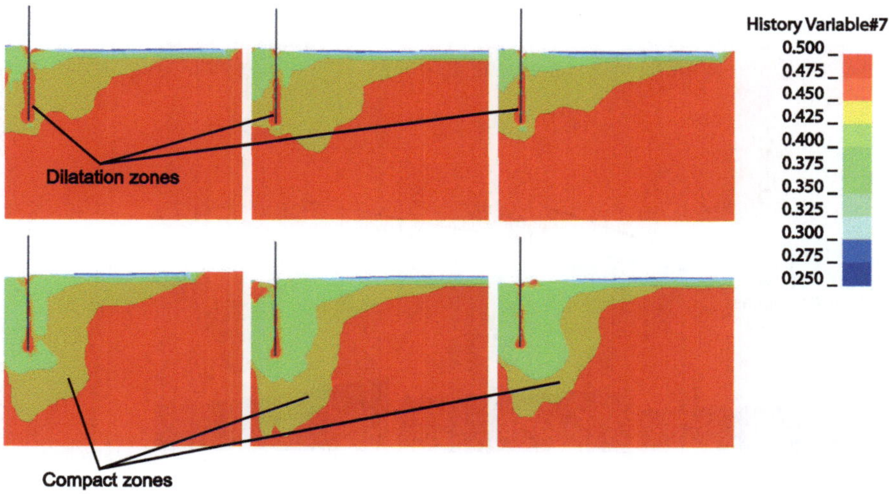

Fig. 7. Void ratio distribution after t = 5 s of pile driving corresponding to different frequencies (initial void ratio, $e_{initial}$ = 0.465)

In the problem under study, this issue is noticed for the model with the frequency of 30, i.e. a less penetration is achieved for the model with the frequency of 30 Hz compared to the model with the frequency of 27 Hz.

Figure 8 shows the lateral stress distribution in the soil after 5 s of pile driving. The lateral stress is important in defining the soil plugging. The investigation of the plugging is important since it contributes to the pile bearing capacity, yet increases soil resistance against pile driving. The influenced area varies according to the pile frequency, i.e. with more frequency, a larger area is affected. The stress distribution near the pile is noisy which can be attributed to the soil disturbance.

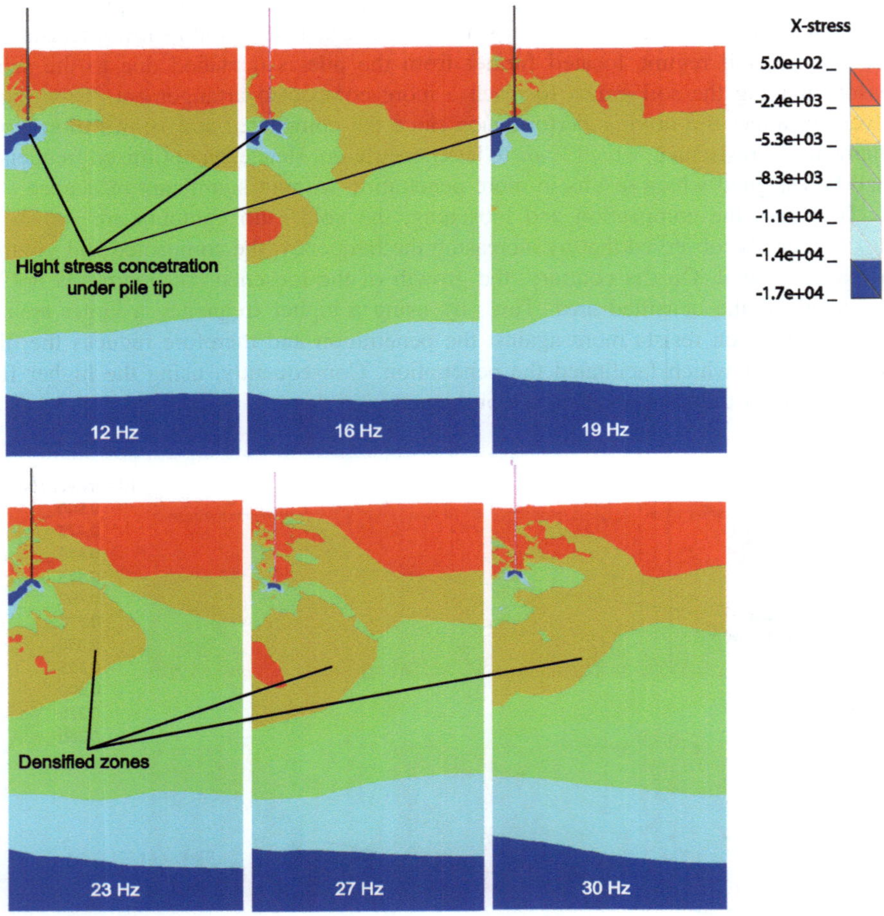

Fig. 8. Horizontal stress distribution after t = 5 s of pile driving corresponding to different frequencies

To study the difference in the lateral stress values inside and outside the pile during driving, the induced lateral forces curves on the pile are plotted in Fig. 9. For higher frequency ranges, a significant increase in stress values inside the pile occurs after about 4 s, which is the similar time stamp as when the differences in penetration curves appeared. This slope change is argued to be related to the plugging initiation (Henke and Bienen 2013). For lower frequency ranges, a small difference is seen, meaning that the soil lateral stress inside and outside the soil are pretty identical.

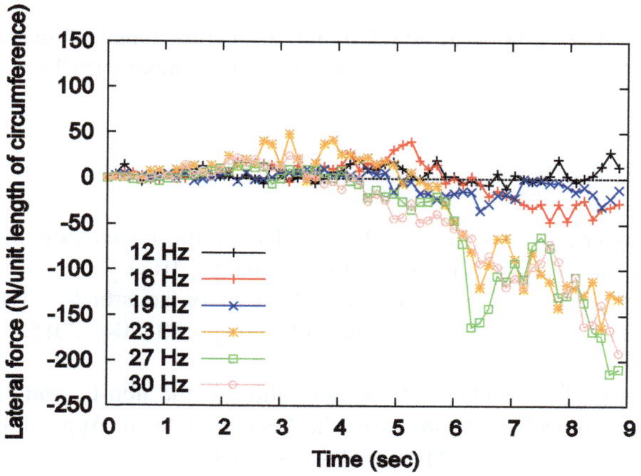

Fig. 9. Total radial force distribution in the whole pile length corresponding to different frequencies (the negative value indicate larger internal than external soil radial stress)

4 Conclusions

In this study, a numerical model was developed using the MMALE method in conjunction with a hypoplastic material model to investigate the effects of the frequency magnitude on the evolution of soil stress and density (void ratio) during the vibratory installation of a tubular pile. The penetration depth of the numerical model was compared to the experimental measurement to validate the model. Afterward, a parametric study was performed where the applied frequency in the vibratory load was changed to study soil behavior.

The resulting penetration depths were first compared. At early stages of the penetration, no significant differences were observed since the static load was dominating, making the frequency load effects negligible. Frequency effects become more apparent at large penetration depths. Due to the vibratory driving, the soil regime near the pile is disturbed, which reduces the soil resistance and thus results in more penetration. In addition, by raising the frequency value, more penetration was observed until reaching the value of 27 Hz. Values more than 27 Hz resulted in less penetration.

To study the underlying reason for less penetration, void ratio and lateral stresses were assessed. It was concluded that by using more frequency, more area is densified, which makes the soil stiffer. Therefore, more resistance is observed against driving. Concerning the simultaneous effect of soil disturbance and the densification on pile driving, it is concluded that an optimum frequency range should be determined to ensure the maximum penetration using the same load magnitude.

The employed numerical model used in this study can be used in further investigations such as pile driving in an undrained condition. Besides, other forms of driving methods can be compared against each other to reach a suitable pile performance.

Acknowledgments. The authors are thankful for the partial financial support obtained from German Academic Exchange Service (DAAD) with grant number 91561676.

References

Aubram, D.: An Arbitrary Lagrangian-Eulerian Method for Penetration into Sand at Finite Deformation. Shaker Verlag, Aachen, Berlin, Germany (2013)

Aubram, D.: Development and experimental validation of an arbitrary Lagrangian-Eulerian (ALE) method for soil mechanics. Geotechnik **38**(3), 193–204 (2015). https://doi.org/10.1002/gete.201400030

Aubram, D., Rackwitz, F., Savidis, S.A.: Contribution to the non-Lagrangian formulation of geotechnical and geomechanical processes. In: Lecture Notes in Applied and Computational Mechanics, vol. 82, pp. 53–100. Springer International Publishing (2017). https://doi.org/10.1007/978-3-319-52590-7_3

Bakroon, M., Daryaei, R., Aubram, D., Rackwitz, F.: Arbitrary Lagrangian-Eulerian finite element formulations applied to geotechnical problems. In: Grabe, J. (ed.) Numerical Methods in Geotechnics, pp. 33–44. BuK! Breitschuh & Kock GmbH, Hamburg, Germany (2017). ISBN: 978-3-936310-43-6

Bakroon, M., Daryaei, R., Aubram, D., Rackwitz, F.: Implementation and validation of an advanced hypoplastic model for granular material behavior. In: 15th International LS-DYNA® Users Conference, p. 12. Detroit, Michigan, USA (2018a)

Bakroon, M., Daryaei, R., Aubram, D., Rackwitz, F.: Multi-material arbitrary Lagrangian-Eulerian and coupled Eulerian-Lagrangian methods for large deformation geotechnical problems. In: Sagaseta, C. (ed.) Numerical Methods in Geotechnical Engineering (NUMGE 2018), p. 8. Porto, Portugal (2018b)

Belytschko, T., Liu, W.K., Moran, B.: Nonlinear Finite Elements for Continua and Structures. Advances in Cancer Research, vol. 104. Wiley, Chichester (2000). https://doi.org/10.1016/s0065-230x(09)04001-9

Benson, D.J.: Computational methods in Lagrangian and Eulerian hydrocodes. Comput. Methods Appl. Mech. Eng. **99**(2–3), 235–394 (1992). https://doi.org/10.1016/0045-7825(92)90042-I

Benson, D.J., Okazawa, S.: Contact in a multi-material Eulerian finite element formulation. Comput. Methods Appl. Mech. Eng. **193**(39–41 spec. iss.), 4277–4298 (2004). https://doi.org/10.1016/j.cma.2003.12.061

De Nicola, A., Randolph, M.F.: The plugging behaviour of driven and jacked piles in sand. Géotechnique **47**(4), 841–856 (1997). https://doi.org/10.1680/geot.1997.47.4.841

Feng, R., Deschamps, R.J.: A study of the factors influencing the penetration and capacity of vibratory driven piles. Soils Found. **40**(3), 43–54 (2000). https://doi.org/10.3208/sandf.40.3_43

Heins, E., Hamann, T., Grabe, J., Hannot, S.: Numerical investigation of the influence of the driving frequency during pile installation of tubular piles. Geotechnik **39**(2), 98–109 (2016). https://doi.org/10.1002/gete.201600014

Henke, S.: Influence of pile installation on adjacent structures. Int. J. Numer. Anal. Methods Geomech. **34**(11), 1191–1210 (2010). https://doi.org/10.1002/nag.859

Henke, S., Bienen, B.: Centrifuge tests investigating the influence of pile cross-section on pile driving resistance of open-ended piles. Int. J. Phys. Model. Geotech. **13**(2), 50–62 (2013). https://doi.org/10.1680/ijpmg.12.00012

Henke, S., Bienen, B.: Investigation of the influence of the installation method on the soil plugging behaviour of a tubular pile. Phys. Model. Geotech. (ICPMG2014). **2**(1995), 681–687 (2014). https://doi.org/10.1201/b16200-94

Henke, S., Grabe, J.: Numerical investigation of soil plugging inside open-ended piles with respect to the installation method. Acta Geotech. **3**(3), 215–223 (2008). https://doi.org/10.1007/s11440-008-0079-7

Holeyman, A., Whenham, V.: Critical review of the Hypervib1 model to assess pile vibro-drivability. Geotech. Geol. Eng. **35**(5), 1933–1951 (2017). https://doi.org/10.1007/s10706-017-0218-8

Kolymbas, D.: A rate-dependent constitutive equation for soils. Mech. Res. Commun. **4**(6), 367–372 (1977). https://doi.org/10.1016/0093-6413(77)90056-8

Le, V.H., Remspecher, F., Rackwitz, F.: Influence of installation effects on the cyclic behaviour of monopile foundation for offshore wind power turbine. In: Randolph, M., Pham, K.H. (eds.) Submitted to "The First Vietnam Symposium on Advances in Offshore Engineering", p. 11, Hanoi, Vietnam (2018)

Niemunis, A., Herle, I.: Hypoplastic model for cohesionless soils with elastic strain range. Mech. Cohesive-frict. Mater. **2**(4), 279–299 (1997)

Qin, Z., Chen, L., Song, C., Sun, L.: Field tests to investigate the penetration rate of piles driven by vibratory installation. Shock and Vibration (2017). https://doi.org/10.1155/2017/7236956

Rausche, F.: Modeling of vibratory pile driving. In: Proceedings of Vibratory Pile Driving and Deep Soil Compaction (TRANSVIB2002), vol. 12 (2002)

Van Leer, B.: Towards the ultimate conservative difference scheme. IV. A new approach to numerical convection. J. Comput. Phys. **23**(3), 276–299 (1977). https://doi.org/10.1016/0021-9991(77)90095-X

von Wolffersdorff, P.-A.: A hypoplastic relation for granular materials with a predefined limit state surface. Mech. Cohesive-Frict. Mater. **1**(3), 251–271 (1996). http://doi.wiley.com/10.1002/%28SICI%291099-1484%28199607%291%3A3%3C251%3A%3AAID-CFM13%3E3.0.CO%3B2-3

Winslow, A.M. Equipotential zoning of two-dimensional meshes (UCRL-7312), United States (1963)

Critical State Theory for Sand with Fines: A DEM Perspective

Nick Barnett, Mizanur Rahman(✉), Rajibul Karim, and Hoang Bao Khoi Nguyen

School of Natural and Built Environments, University of South Australia, Adelaide, SA, Australia
{Nick.Barnett, Mizanur.Rahman, Rajibul.Karim, Khoi.Nguyen}@unisa.edu.au

Abstract. Undrained triaxial simulations were carried out for three-dimensional assemblies of clean sand and sand with fines soils using the discrete element method (DEM). Similar to experimental observations, it was found that the critical state line (CSL) moved downward with increasing fines content (f_c) in the e-log(p') space; where e is the void ratio and p' is the mean effective confining stress. The CSL for both clean sand and sand with fines was observed to provide an appropriate reference line to predict the dilative tendencies throughout undrained shearing. Soil elements with an initial state (i.e. e and p' at the beginning of undrained shearing) located above the CSL were shown to exhibit contractant behaviour, whilst soil with initial states below the CSL yielded dilatant behaviour. However, this study observed that the classical e is not an appropriate density index for sand with fines. Instead, the equivalent granular void ratio, e^* provided a more suitable density index of sand with fines soils, as it reflects the force skeleton structure for both clean sand and sand with fines.

1 Introduction

The critical state soil mechanics (CSSM) framework has been extensively investigated and is well understood for clean sand behaviour. One of many advantages of the CSSM framework is, if the state of sand, i.e. e and p', are known with respect to the critical state line (CSL) in the e-log(p') space, then its behaviour can be predicted; where e is the void ratio and p' is the mean effective confining stress. Experimental studies have suggested that sands which exhibit a state above the CSL will manifest contractant (liquefiable) behaviour and sands with a state below the CSL display dilatant (non-liquefiable) behaviour. Been and Jefferies (1985) provided a mathematical formulation of such states relative to the CSL, termed the state parameter, ψ, which the difference in e at the current state and on the CSL at the same p'. Hence, a sand with a positive ψ will display contractant behaviour and a sand with a negative ψ will display dilatant behaviour. Furthermore, ψ correlates well with other characteristic responses of granular materials, such as instability behaviour (Goudarzy et al. 2017; Rabbi et al. 2018) and excess pore water pressure (Zhang et al. 2018). While this is elegant, a unified CSSM framework may not be applicable for many natural sands as they often

H. Shehata and C. S. Desai (Eds.): GeoMEast 2018, SUCI, pp. 62–75, 2019.
https://doi.org/10.1007/978-3-030-01926-6_4

contain a small amount of fines content (f_c) i.e. particle sizes smaller than 0.075 mm. It was found in most earlier studies that the CSLs of these sands shift downwards in the e-log(p') space with increasing f_c up to a threshold fines content (f_{thre}) (Ni et al. 2004; Rahman and Lo 2014; Rahman et al. 2008; Thevanayagam et al. 2002; Zlatovic and Ishihara 1997). This is problematic because of the fact that the CSL varies for each f_c and all possible CSLs must be known to evaluate soil behaviour under CSSM framework. Therefore, the development of a unified CSSM framework irrespective of f_c for these sands is a challenge.

Many argued that for the same e, clean sands and sand mixed with f_c do not share the same force skeleton structure, hence e may not be an appropriate state indicator for sand with f_c. In an attempt to capture the effect of the f_c on critical state (CS) behaviour of sands, the concept of equivalent granular void ratio (e*) was introduced (Georgiannou et al. 1990; Ovando-Shelley and Pérez 1997; Thevanayagam and Mohan 2000). The e* in conjunction with a b parameter coalesced the CSLs in e-log(p') space to a single trendline in the e*-log(p') space, irrespective of f_c (Lashkari 2014; Ni et al. 2004; Qadimi and Mohammadi 2014; Rahman and Lo 2008; Rahman et al. 2011; Thevanayagam et al. 2002; Yang et al. 2006). The single trend of CSLs is often referred to as the equivalent granular critical state line (EG-CSL). The EG-CSL has been used to modify ψ to the equivalent granular state parameter (ψ*) (Rahman and Lo 2007). Recent studies have observed that ψ* correlates with the mechanical behaviour of sand with fines (Mohammadi and Qadimi 2015; Rahman et al. 2014). Such observations have led towards the development of state-dependant constitutive modelling applications of sand with fines (Lashkari 2014; Rahman et al. 2014). The conversion of e to e* forms a modified version of the critical state theory (CST) which is now known as equivalent state theory (EST). The underlying assumption of the EST is that the b parameter represents the active fraction of f_c in the force skeleton structure of sand in an idealised sand-fines matrix. Although this assumption is evaluated through experimental approaches (e.g. scanning electron microscopy (SEM), computed tomography (CT) scanning etc.) for sand mixed with f_c, little attention has been paid to its numerical evaluation, especially from a micromechanical perspective.

Cundall and Strack (1979) first proposed DEM and since then it has been increasingly used in an attempt to better understand the complex behaviour of granular materials. Recently, it has been used to evaluate the CS behaviour in clean sands (Gu et al. 2014; Huang et al. 2014; Nguyen and Rahman 2017; Nguyen et al. 2017; Rahman et al. 2018; Zhao and Guo 2013). However, DEM studies evaluating the CST for sand mixed with fines are relatively scarce. It is worth noting that there are few discrete element method (DEM) studies on sand with fines, however the applicability of EST was not assessed (Dai et al. 2015; Gong and Liu 2017; Minh et al. 2014; Ng et al. 2016). Therefore, the objective of this study is to evaluate the CST for sand with fines soils and then assess the appropriateness of e* through a discrete element method (DEM) perspective.

2 Methodology

2.1 DEM Testing Program

This study used a three-dimensional DEM software, PFC^{3D} (Itasca 2008) to undertake a number of undrained triaxial compression simulations on both clean sand and sand with 10% fines (i.e. $f_c = 0.1$) samples. Both sand and fine particles were spherical. The details of the simulation regime are shown in Table 1.

Table 1. Testing program details of DEM simulations

Material	Test ID	p'_0 (kPa)	e_0	N_s	N_f	N
Clean sand ($f_c = 0$)	UD-T2	96	0.597	10,506	–	10,506
	UD-T4	96	0.667	10,506	–	10,506
	UD-T6	96	0.852	10,506	–	10,506
	UD-T7	96	0.796	10,506	–	10,506
	UD-T13	96	0.875	10,506	–	10,506
Sand with fines ($f_c = 0.1$)	UD-T307	94	0.546	2,524	126,742	129,266
	UD-T309	94	0.711	2,524	126,742	129,266
	UD-T311	94	0.678	2,524	126,742	129,266
	UD-T312	94	0.746	2,524	126,742	129,266

*Note: N is total number of particles; N_s is number of sand particles; N_f is number of fine particles.

Both clean sand (Fig. 1a) and sand with monodisperse fines (Fig. 1b) assemblies were generated within six frictionless and rigid walls. Since uncharacteristically high friction angles can be mobilized in smaller assemblies due to the rigid boundary end-restraint, the accuracy of the simulations with rigid walls depends on specimen size

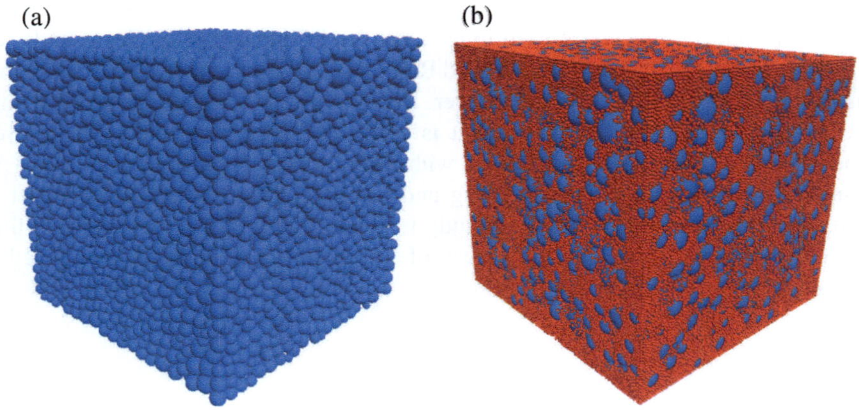

Fig. 1. DEM specimens: (a) clean sand (b) sand with fines ($f_c = 0.1$)

(Huang et al. 2014; Zhou et al. 2017). To minimize the effect of rigid boundaries, a large specimen consisting of a large number of particles needed to be considered at the expense of increased computational demand. This study deployed a trial and error approach to optimise the number of particles for clean sand and sand with 10% fines, so that the accuracy of simulations was not affected by specimen size/number of particles. The process of achieving numerical stability and model optimisation is further discussed in subsequent sections.

The particle size distribution (PSD) of both materials are illustrated in Fig. 2. The sand assembly has a PSD identical to Sydney sand. The size of the monodisperse fines were prepared in such a way that particle size ratio $\chi = D/d \approx 6.5$; where D and d are size of the sand and fines particles respectively. $\chi \geq 6.5$ is a geometric requirement of particle size ratio which allows fines to be trapped in within the densest possible packing of larger particles and to achieve a lower theoretical density (Rahman et al. 2008).

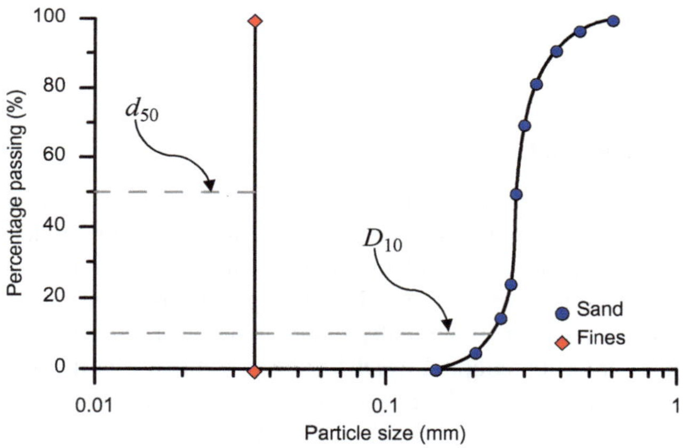

Fig. 2. Particle size distribution of Sydney sand and fines

A staged isotropic compression method was used to achieve a wide range of initial densities (i.e. from loose to dense) for DEM assemblies. The assembly was first generated with no initial contact. A small isotropic stress was applied to the soil via a servo-control mechanism applied to each wall. Throughout this phase of isotropic compression, the inter-particle sliding friction coefficient (μ) is set between zero and one depending on the desired end density. $\mu = 0$ produces a very dense assembly, and $\mu = 1$ produces a very loose assembly. Intermediate values of μ were used to generate mid-range densities. Once a prescribed isotropic stress and a desirable initial void ratio (e_0) is achieved, μ is set to 0.5 for the triaxial simulation. The triaxial simulation begins when the specimen commences further isotropic consolidation to a prescribed p'_0 value. On completion of consolidation, undrained shearing proceeded. Undrained shearing was achieved by gradual strain control of the top and bottom walls, whilst the no volume change condition was maintained through another servo-control mechanism.

Table 2 specifies the input parameters used for the triaxial simulations. These parameters are reflective of both the input parameters for sand and fines materials, and are also commonly adhered to in literature (Gu et al. 2014).

Table 2. Input parameters used in this study

Parameter	Value
Particle density (ρ_g)	2650 kg/m^3
Effective Young's modulus (E*)	50 MPa
Interparticle sliding friction coefficient (μ)	0.5
Interparticle rolling resistance coefficient (μ_r)	0.3
Wall friction coefficient (μ_{wall})	0
Particle normal contact stiffness (k_n)	Variable (as per Eq. 1)
Particle tangential contact stiffness (k_s)	$k_s = k_n$
Particle rolling contact stiffness (k_s)	Variable (as per Eq. 1)
Wall stiffness (k_{wall})	1×10^{10} kN/m
Damping constant	0.7

2.2 Contact Model and Input Parameters

A rolling resistance linear contact model (RRLCM) was used in this study. The RRLCM is similar to the commonly used linear contact model but with the inclusion of a rolling resistance mechanism. Soils in nature commonly possess some angularity and surface texture features. However, generating such angular particles in DEM is challenging and computationally demanding. Therefore, idealised circular or spherical particles are often adopted in DEM to represent soil particles. This may lead to particles developing large rolling tendencies throughout shearing, which has the capacity to influence overall soil behaviour in simulation (Ai et al. 2011). This may specially be the case for sand with fines soils, where the large particle size ratio provides an increased likelihood for rolling to prevail (Dai et al. 2015). Therefore, a rolling resistance coefficient (μ_r) is introduced to allow idealised (spherical) DEM particles to mimic the rolling resistance behaviour of realistic angular soil particles. Consistent with a previous DEM study for sand with fines by Gong and Liu (2017), a μ_r value of 0.3 for both sand and fine particles was used in this study.

The adopted rolling resistance linear contact model computes normal contact stiffness (k_n), tangential contact stiffness (k_s) and rolling contact stiffness (k_r) between two contacting entities based on the following relationships:

$$k_n = \frac{AE^*}{L}; k_s = \frac{k_n}{k^*}; k_r = k_s \bar{R}^2; \qquad (1)$$

where E* is the effective Young's modulus, A is the area of the smallest particle in contact, L is the contact length as illustrated in Fig. 3, k* is the ratio between k_n and k_s and $\bar{R}^2$ is the effective radius between two contacting entities. Based on the stiffness

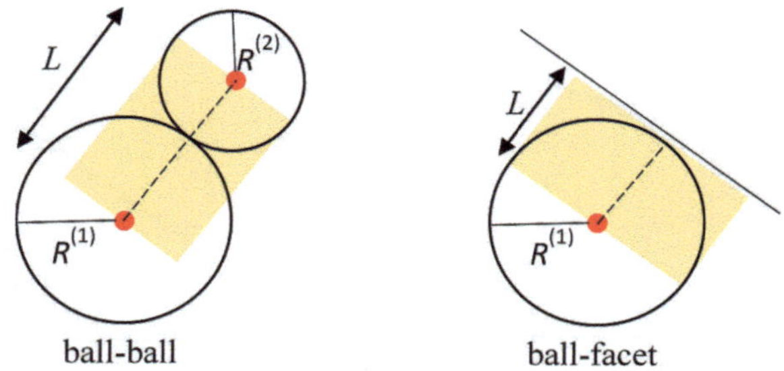

Fig. 3. Conceptualisation of contact length in PFC^{3D}

calculations between two contacting pieces, a force-displacement law was used interchangeably with Newton's Second Law to compute inter-particle contact forces and particle displacements at each small time increment.

2.3 Numerical Stability of DEM Simulation

To ensure the modelled granular assembly exhibits a realistic behaviour under monotonic loading, conditions for quasi-static flow were assessed and met. The inertial number (*I*) during shearing is a common criteria used to assess the quasi-static flow condition, which was defined by da Cruz et al. (2005) as:

$$I = \dot{\varepsilon} d \left(\rho_g / p'\right)^{0.5} \tag{2}$$

where $\dot{\varepsilon}$ is the strain rate, d is the mean particle diameter and ρ_g is the average particle density. A small value of I ($\leq$ 0.003) indicates the inertial effects of individual particles are small and have negligible influence on the overall (static) behaviour of the soil (da Cruz et al. 2005; Nguyen et al. 2018). This condition was satisfied throughout all simulations reported in this paper.

Assessing the sensitivity of the chosen shear deformation rate ($d\varepsilon_q$) throughout shearing may also be undertaken to evaluate numerical stability. Nguyen et al. (2017) suggested $d\varepsilon_q$ must be small enough to ensure a quasi-static progression of particle rearrangements. Furthermore, Andrade et al. (2012) suggested that the unwise selection of $d\varepsilon_q$ may have a profound effect on the CS response of granular material. $d\varepsilon_q = 0.002$ was used in this study. Figure 4 shows no observation of sensitivity (in q-ε_q space and q-p' space) between simulations conducted with $d\varepsilon_q = 0.002$/s and $d\varepsilon_q = 0.0004$/s (5 times smaller).

In DEM simulations, it is important to ensure the yielded results are independent of number of particles (*N*) and any potential boundary effect (Ng 2004; Nguyen et al. 2017). Figure 5 depicts the influence of number of particles on soil behaviour for sand with 10% fines (N = 129,266 and N = 144,506). Some discrepancy in behaviour may be attributed to the differences in e, as also suspected by Nguyen et al. (2017). It is pertinent to note, that in DEM, reproducing the same values of e in different simulations is challenging and time consuming. As per Fig. 5, size independence was

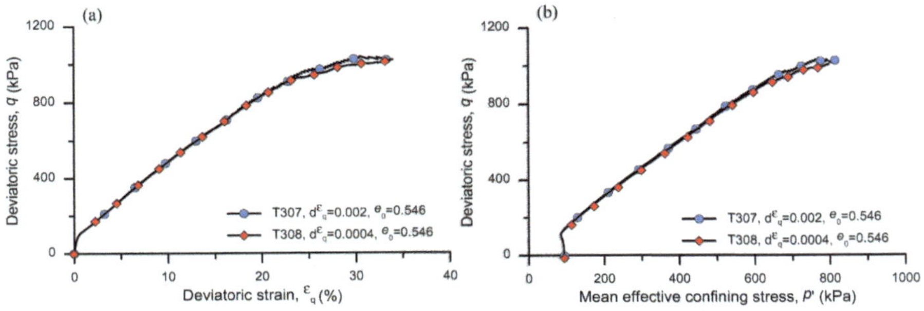

Fig. 4. Effect of strain rate on overall behaviour of sand with fines (f_c = 0.10): (a) in q-ε_q space (b) in q-p' space.

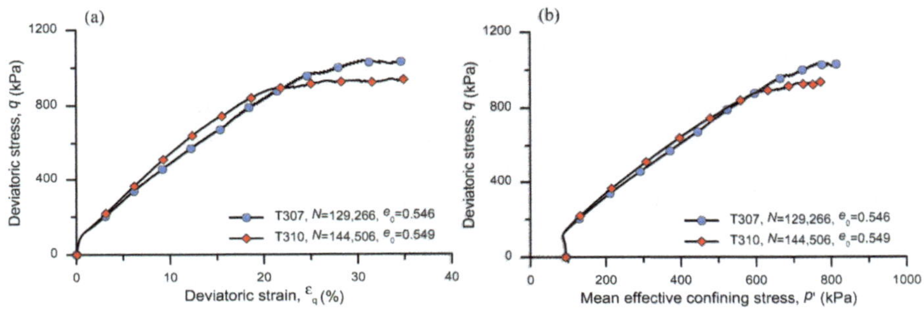

Fig. 5. Effect of number of particles on overall behaviour of sand with fines (f_c = 0.10): (a) in q-ε_q space (b) in q-p' space.

achieved and therefore N = 129,266 was used in simulation for the sand with 10% fines soil mixture. Identical analyses were also undertaken for the clean sand assembly.

Additionally, PFC^{3D} code was developed in a way that simulation stages (i.e. specimen generation, staged isotropic compression and consolidation) can only be completed once an unbalanced force ratio which corresponds to quasi-static equilibrium is achieved.

3 Results

Undrained simulation results for clean sand and sand with fines samples are discussed here and illustrated in Fig. 6, 7 and 8. The deviatoric stress $\left(q = \sigma_1' - \sigma_3'\right)$ and mean effective confinement stress $\left(p' = \left[\sigma_1' + 2\sigma_3'\right]/3\right)$ were computed using the stress tensor developed by Christoffersen et al. (1981):

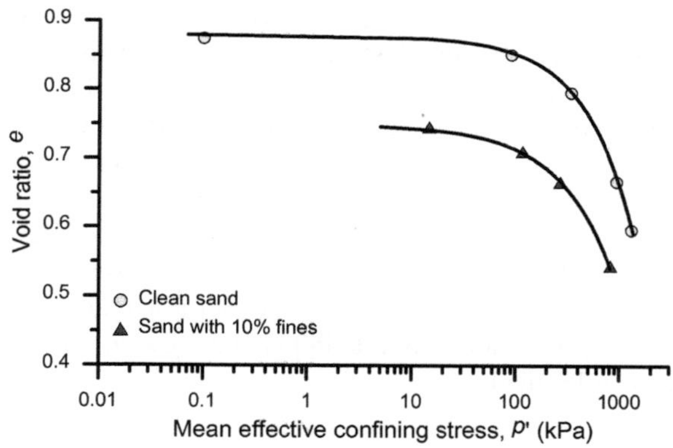

Fig. 6. CSLs for simulated granular materials

$$\sigma'_{ij} = \frac{1}{V} \sum_{c \in N_c} f_i^c l_j^c \tag{3}$$

where V is the total assembly volume, N_c is the total number of contacts, f_i^c represents the ith component contact force at contact c and l_j^c represents the jth branch vector connecting the centres of two contacting particles, at contact c. Commonly in DEM, the soil stresses are computed by considering the contact forces of particles acting upon the rigid boundaries. However, to reduce the likelihood of rigid boundaries expressing artificial behaviour, the stress tensor which considers only internal contact forces was utilised.

3.1 Critical State Line for Sand with Fines Soils

Based on the undrained simulations for clean sand and sand with fines (as per Table 1), a CSL for each granular assembly was developed in e-log p' space (Fig. 6). As fines were added to the host sand, a downward shift of the CSL occurred. Thus, with the addition of a small amount of fines, the CS strength of the soil deteriorates, i.e. lower CS strength at the same e. The findings, as shown in Fig. 6, are consistent with many CS experimental studies (Rahman et al. 2011; Thevanayagam et al. 2002).

3.2 Critical State Line (CSL) Projecting Soil Behaviour

Experimental studies have established that the traditional CSL can be effectively used to predict the dilative and liquefaction behaviour of clean sand. In this study, the CSLs ability to predict undrained behaviour of both clean sand and sand with fines soils (e.g. contractant, contractant-dilatant and dilatant) in terms of the material initial state (i.e. e and p' at the beginning of shearing) was assessed. First, in Fig. 7a, the initial states of clean sand specimens located above the respective CSL were shown to exhibit contractant behaviour throughout undrained shearing, whereas the specimens which possessed an initial state below the CSL displayed dilatant behaviour. On inspection of Fig. 7a, one specimen had an initial state on the CSL. This specimen manifested contractant-dilatant (transitional) behaviour. In Fig. 7b, the initial states of sand with 10% fines are compared with the respective CSL. Similar observations are made here i.e. an initial state above the CSL exhibit contractant behaviour and below the CSL specimens exhibited dilatant behaviour. This is consistent with the experimental observation for Sydney sand with fines (Rahman and Lo 2014; Rahman et al. 2011). Therefore, the CSLs for clean sand and sand with fines obtained from DEM also provide a convenient reference line which can predict granular material behaviour i.e. contractant and dilatant behaviour from its initial state. Based on Fig. 7, the following three observations can be made,

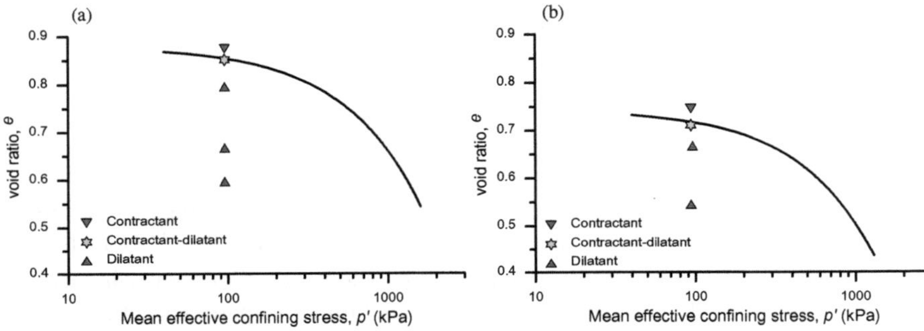

Fig. 7. Influence of initial states with respect to the CSL on the types of undrained behaviour: (a) clean sand; (b) sand with fines (f_c = 0.10).

1. The CSL for both clean sand and sand with fines soils becomes an appropriate reference line to predict the soil states associated with liquefiable or non-liquefiable behaviour,
2. Through knowledge of the soils initial state, the contractant, dilatant and contractant-dilatant tendencies of clean sand and sand with fines soils can be predicted,
3. The CSL can capture identical behavioural aspects for sand with fines soils the same way it does for clean sand.

3.3 Equivalent Granular Void Ratio e^*

Although the CSL in e-log p' space is useful in predicting the liquefaction behaviour of sand with fines soils, using classical CSSM framework to analyse its behaviour becomes somewhat impractical as a separate CSL is needed for each f_c. As a result, the equivalent granular void ratio (e^*) was proposed by Thevanayagam et al. (2002), which later evolved as EST for constitutive modelling (Lashkari 2016; Rahman et al. 2014). In EST, e is replaced with a new density index for sand with fines, namely, e^*. Conveniently, e^* coalesces the CSLs (in e-log p' space) of sand with fines soils up to a threshold fines content, f_{thre} to one single CSL, termed the equivalent granular critical state line (EG-CSL) which is projected in e^*-log p' space. Therefore, through the application of CSSM framework, e^* becomes a preferable density index for sand with fines soils. Further, experimental studies suggest that e^* provides a more realistic density index than the classical e (Rahman et al. 2014). e^* was mathematically defined by Thevanayagam et al. (2002) as:

$$e^* = \frac{e + (1 - b)f_c}{1 - (1 - b)f_c} \tag{4}$$

where the parameter b, captures the active fraction of fines in the force skeleton structure and is computed using the equation below (after Rahman and Lo (2008); Rahman et al. (2009)).

$$b = \left[1 - \exp\left(-0.3\frac{(f_c/f_{thre})}{k}\right)\right] \times \left(r\frac{f_c}{f_{thre}}\right)^r \tag{5}$$

where, $r = 1/\chi = d/D$ and $k = 1 - r^{0.25}$. Since sand and fines are generally not single-sized materials, D/d was generalized to D_{10}/d_{50} based on the argument in Ni et al. (2004), where the subscripts denote the fractile passing. In this study, $D_{10} = 0.23$ mm and $d_{50} = 0.0354$ mm (see Fig. 2). f_{thre} may be obtained from experimental data, where available, as outlined in Rahman et al. (2009). Alternatively, as an initial approximation, f_{thre} can be taken as 0.30, or it may be determined more reliably using the following equation developed by Rahman et al. (2009).

$$f_{thre} = 0.40\left(\frac{1}{1 + \exp^{\alpha - \beta\chi}} + \frac{1}{\chi}\right) \tag{6}$$

The parameters α and β have been determined by curve fitting of eight datasets for χ in the range of 2 to 42, and this gave $\alpha = 0.50$ and $\beta = 0.13$.

3.4 Evaluation of the use of e^*

e has been successfully used as a density index to capture the CS response for clean sands which has resulted in the effective use of CST for clean sands. However, as highlighted previously when the same host sand is mixed with a small amount of fines, the value of e may not be representative in explaining its behaviour. For example, a clean sand and a sand with 10% fines specimen both undergoing undrained shearing at similar values of e and p' values may exhibit dissimilar behaviours as shown in Fig. 8. By the definition of e, the sand with fines mixture is denser than the clean sand soil, however the denser soil shows overall less dilative behaviour and displays a smaller CS strength. Such a finding contradicts CST and fundamental engineering mechanics, suggesting that the classical e is not an appropriate density index for both clean sand and sand with fines soils. The concept of e^* provides a more appropriate density index for sand with fines. When the density index is expressed in terms of e^* for T307 (as per Eq. 4, see Fig. 8), the deviatoric stress-strain response in comparison to the clean sand becomes more understandable (i.e. the denser soil displays more dilative behaviour).

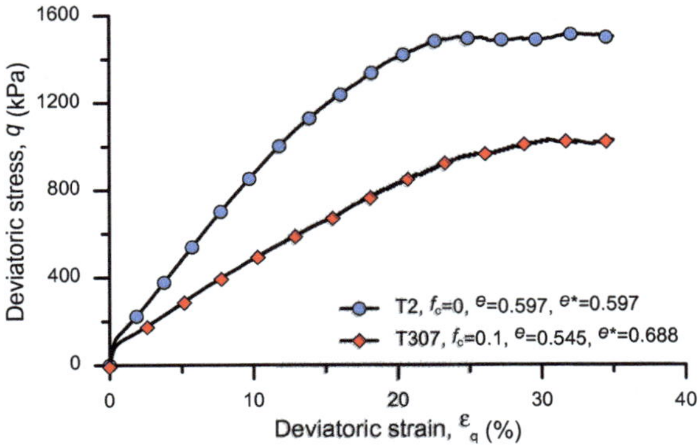

Fig. 8. Comparison of e and e^* as state index for clean sand and sand with fines

Thus, e^* better represents the force skeleton structure of the soil, in comparison to e. The observations therefore suggest that sand with fines behaviour may be better represented through the EST rather than the classical CST.

4 Conclusions

The CST and applicability of e^* for sand with fines soils has been investigated through DEM. The main findings and conclusions drawn from this numerical study are as follows:

- Achieving numerical stability in three-dimensional DEM simulations for sand with fines soils may be challenging due to the large number of particles needed to simulate a small volume. Although, this study successfully maps out a regime, in which computational expenses can be minimised, whilst achieving numerical stability.
- The CSL (in e-log p' space) for both clean sand and sand with fines is observed to be an accurate tool to predict the soil states associated with contractant, dilatant and contractant-dilatant behaviour. However, a separate CSL is needed for each f_c, which is an impractical and laborious task for granular material with a range of f_c.
- The observed behaviour for clean sand and sand with fines soils was not reflected by the classical density index, e, alone. Instead, e^* better reflects the force skeleton structure of sand with fines soils and hence provides a more appropriate density index for both clean sand and sand with fines.

The study used DEM assessed the effect of fines on undrained behaviour of coarse granular material under CSSM framework. Although DEM provides an ideal platform to analyse the influence of fine particle, the reader should not directly extrapolate these understanding to real soil behaviour.

Acknowledgements. The first author would like to acknowledge the support provided by the Research Training Program domestic (RTPd) scholarship awarded by The School of Natural and Built Environments, University of South Australia.

References

Ai, J., Chen, J., Rotter, M., Ooi, J.Y.: Assessment of rolling resistance models in discrete element simulations. Powder Technol. **206**, 269–282 (2011)

Andrade, J., Chen, Q., Le, P., Avila, C., Evans, T.: On the rheology of dilative granular media: bridging solid- and fluid-like behavior. J. Mech. Phys. Solids **60**, 1122–1136 (2012)

Been, K., Jefferies, M.G.: A state parameter for sands. Geotechnique **35**(2), 99–112 (1985)

Christoffersen, J., Mehrabadi, M.M., Nemat-Nasser, S.: A micromechanical description of granular material behavior. J. Appl. Mech. **48**(2), 339–344 (1981). https://doi.org/10.1115/1.3157619

Cundall, P.A., Strack, O.D.L.: A discrete numerical model for granular assemblies. Geotechnique **29**(1), 47–65 (1979). https://doi.org/10.1680/geot.1979.29.1.47

da Cruz, F., Emam, S., Prochnow, M., Roux, J.-N., Chevoir, F.: Rheophysics of dense granular materials: discrete simulation of plane shear flows. Phys. Rev. E **72**(2), 021309 (2005). https://doi.org/10.1103/PhysRevE.72.021309

Dai, B., Yang, J., Luo, X.: A numerical analysis of the shear behavior of granular soil with fines. Particuology **21**, 160–172 (2015)

Georgiannou, V.N., Burland, J.B., Hight, D.W.: The undrained behaviour of clayey sands in triaxial compression and extension. Geotechnique **40**(3), 431–449 (1990)

Gong, J., Liu, J.: Mechanical transitional behavior of binary mixtures via DEM: effect of differences in contact-type friction coefficients. Comput. Geotech. **85**, 1–14 (2017). https://doi.org/10.1016/j.compgeo.2016.12.009

Goudarzy, M., Rahemi, N., Rahman, M.M., Schanz, T.: Predicting the maximum shear modulus of sands containing nonplastic fines. J. Geotech. Geoenviron. Eng. **143**(9), 06017013 (2017). https://doi.org/10.1061/(asce)gt.1943-5606.0001760

Gu, X., Huang, M., Qian, J.: DEM investigation on the evolution of microstructure in granular soils under shearing. Granul. Matter **16**(1), 91–106 (2014). https://doi.org/10.1007/s10035-013-0467-z

Huang, X., Hanley, K.J., O'Sullivan, C., Kwok, F.C.: Effect of sample size on the response of DEM samples with a realistic grading. Particuology **15**, 107–115 (2014a)

Huang, X., O'Sullivan, C., Hanley, K., Kwok, C.: Discrete-element method analysis of the state parameter. Geotechnique **64**(12), 954–965 (2014b)

Itasca. Particle Flow Code in 3 Dimensions (PFC3D) version 4, Minnesota, USA (2008)

Lashkari, A.: Recommendations for extension and re-calibration of an existing sand constitutive model taking into account varying non-plastic fines content. Soil Dyn. Earthq. Eng. **61–62**, 212–238 (2014). https://doi.org/10.1016/j.soildyn.2014.02.012

Lashkari, A.: Prediction of flow liquefaction instability of clean and silty sands. Acta Geotech. **11**(5), 987–1014 (2016). https://doi.org/10.1007/s11440-015-0413-9

Minh, N., Cheng, Y., Thornton, C.: Strong force networks in granular mixtures. Granul. Matter **16**(1), 69–78 (2014)

Mohammadi, A., Qadimi, A.: A simple critical state approach to predicting the cyclic and monotonic response of sands with different fines contents using the equivalent intergranular void ratio. Acta Geotech. **10**(5), 587–606 (2015). https://doi.org/10.1007/s11440-014-0318-z

Ng, T.-T., Zhou, W., Chang, X.-L.: Effect of particle shape and fine content on the behavior of binary mixture. J. Eng. Mech. **143**(1) (2016). https://doi.org/10.1061/(asce)em.1943-7889.0001070

Ng, T.: Triaxial test simulations with discrete element method and hydrostatic boundaries. J. Eng. Mech. **130**(10), 1188–1194 (2004). https://doi.org/10.1061/(asce)0733-9399(2004)130:10(1188)

Nguyen, H.B.K., Rahman, M.M.: The role of micro-mechanics on the consolidation history of granular materials. Australian Geomech. **52**(3), 27–36 (2017)

Nguyen, H.B.K., Rahman, M.M., Fourie, A.B.: Undrained behaviour of granular material and the role of fabric in isotropic and K0 consolidations: DEM approach. Geotechnique **67**(2), 153–167 (2017). https://doi.org/10.1680/jgeot.15.p.234

Nguyen, H.B.K., Rahman, M.M., Fourie, A.B.: Characteristic behaviour of drained and undrained triaxial tests: a DEM study. J. Geotech. Geoenviron. Eng. (In press) (2018). https://doi.org/10.1061/(ASCE)GT.1943-5606.0001940

Ni, Q., Tan, T.S., Dasari, G.R., Hight, D.W.: Contribution of fines to the compressive strength of mixed soils. Geotechnique **54**(9), 561–569 (2004)

Ovando-Shelley, E., Pérez, B.E.: Undrained behaviour of clayey sands in load controlled triaxial tests. Geotechnique **47**(1), 97–111 (1997)

Qadimi, A., Mohammadi, A.: Evaluation of state indices in predicting the cyclic and monotonic strength of sands with different fines contents. Soil Dyn. Earthq. Eng. **66**, 443–458 (2014). https://doi.org/10.1016/j.soildyn.2014.08.002

Rabbi, A.T.M.Z., Rahman, M.M., Cameron, D.A.: Undrained behavior of silty sand and the role of isotropic and K0 consolidation. J. Geotech. Geoenviron. Eng. **144**(4), 1–11 (2018). https://doi.org/10.1061/(ASCE)GT.1943-5606.0001859

Rahman, M.M., Lo, S.-C.R., Dafalias, Y.F.: Modelling the static liquefaction of sand with low-plasticity fines. Geotechnique **64**(11), 881–894 (2014). https://doi.org/10.1680/geot.14.P.079

Rahman, M.M., Lo, S.R.: Equivalent granular void ratio and state parameters for loose clean sand with small amount of fines. In: 10th Australia New Zealand Conference on Geomechanics: Common Ground, pp. 674–679 (2007)

Rahman, M.M., Lo, S.R.: The prediction of equivalent granular steady state line of loose sand with fines. Geomech. Geoeng. **3**(3), 179–190 (2008). https://doi.org/10.1080/17486020802206867

Rahman, M.M., Lo, S.R.: Undrained behaviour of sand-fines mixtures and their state parameters. J. Geotech. Geoenviron. Eng. **140**(7) (2014). https://doi.org/10.1061/(asce)gt.1943-5606.0001115

Rahman, M.M., Lo, S.R., Baki, M.A.L.: Equivalent granular state parameter and undrained behaviour of sand-fines mixtures. Acta Geotech. **6**(4), 183–194 (2011). https://doi.org/10.1007/s11440-011-0145-4

Rahman, M.M., Lo, S.R., Gnanendran, C.T.: On equivalent granular void ratio and steady state behaviour of loose sand with fines. Can. Geotech. J. **45**(10), 1439–1455 (2008). https://doi.org/10.1139/T08-064

Rahman, M.M., Lo, S.R., Gnanendran, C.T.: Reply to discussion by Wanatowski, D. and Chu, J. on—On equivalent granular void ratio and steady state behaviour of loose sand with fines. Can. Geotech. J. **46**(4), 483–486 (2009). https://doi.org/10.1139/T09-025

Rahman, M.M., Nguyen, H.B.K., Rabbi, A.T.M.Z.: The effect of consolidation on undrained behaviour of granular materials: experiment and DEM simulation. Geotech. Res. **5**(3), 1–19 (2018). https://doi.org/10.1680/jgere.17.00019

Thevanayagam, S., Mohan, S.: Intergranular state variables and stress-strain behaviour of silty sands. Geotechnique **50**(1), 1–23 (2000)

Thevanayagam, S., Shenthan, T., Mohan, S., Liang, J.: Undrained fragility of clean sands, silty sands, and sandy silts. J. Geotech. Geoenviron. Eng. **128**(10), 849–859 (2002)

Yang, S.L., Sandven, R., Grande, L.: Steady-state lines of sand-silt mixtures. Can. Geotech. J. **43** (11), 1213–1219 (2006)

Zhang, J., Lo, S.-C.R., Rahman, M.M., Yan, J.: Characterizing monotonic behavior of pond ash within critical state approach. J. Geotech. Geoenviron. Eng. **144**(1), 04017100 (2018). https://doi.org/10.1061/(ASCE)GT.1943-5606.0001798

Zhao, J., Guo, N.: Unique critical state characteristics in granular media considering fabric anisotropy. Geotechnique 695–704 (2013)

Zhou, W., Wu, W., Ma, G., Huang, Y., Chang, X.: Study of the effects of anisotropic consolidation on granular materials under complex stress paths using the DEM. Granul. Matter **19**(4), 76 (2017). https://doi.org/10.1007/s10035-017-0763-0

Zlatovic, S., Ishihara, K.: Normalized behaviour of very loose non-plastic soils: effect of fabric. Soils Found. **37**(4), 47–56 (1997)

Modelling the Liquefaction Behaviour of Sydney Sand and the Link Between Static and Cyclic Instability

Md. Mizanur Rahman[1(✉)], Hoang Bao Khoi Nguyen[2], and Md. Rajibul Karim[2]

[1] Geotechnical Engineering, School of Natural and Built Environments, University of South Australia, Adelaide, Australia
Mizanur.Rahman@unisa.edu.au

[2] School of Natural and Built Environments, University of South Australia, Adelaide, Australia
{Khoi.Nguyen, Rajibul.Karim}@unisa.edu.au

Abstract. A state dependent constitutive model, from the family of SANI-SAND models, was adopted to develop a constitutive model for Sydney sand to predict static liquefaction behaviour and to simulate cyclic instability behaviour. The detail formulation of the constitutive model within the critical state soil mechanics (CSSM) framework was presented first. The model was then used to predict a static liquefaction behaviour which showed good match for characteristic behaviour of undrained tests. The model was also used to evaluate the link between static and cyclic instability behaviour. Although there were some limitations in the model simulation, they shows a good agreement with the literature that static liquefaction and cyclic instability are linked through the instability zone.

1 Introduction

Three types on undrained monotonic behaviours are commonly observed, namely, flow liquefaction, limited flow liquefaction and non-liquefaction. These observations depend on the state of sand, i.e. void ratio (*e*) and mean effective confining stress (p'). Both flow liquefaction and limited flow liquefaction manifest instability behaviour where static liquefaction triggers. The instability in the context of continuum mechanics, i.e. $d\sigma_{ij}d\varepsilon_{ij} < 0$ (Lade 1994), is manifested, where σ_{ij} and ε_{ij} are stress and strain tensors respectively and "d" represents infinitesimal increments. Similarly, depending on the state of sand, two different types of cyclic liquefaction often are observed: cyclic instability and cyclic mobility. Cyclic instability involves abrupt run-off deformation in conjunction with rapid pore water pressure generation in loose saturated sand, which is similar to flow liquefaction due to monotonic loading. Upon triggering cyclic instability, sand loses its strength significantly within few loading cycles (Baki et al. 2012; Vaid et al. 2001; Yamamuro and Covert 2001; Yang and Sze 2011). Therefore, the reduction of strength persist after cessation of cyclic loading (provided re-consolidation has not occurred). It should be noted that many different terminologies were used in the

H. Shehata and C. S. Desai (Eds.): GeoMEast 2018, SUCI, pp. 76–86, 2019.
https://doi.org/10.1007/978-3-030-01926-6_5

literature for this form of cyclic liquefaction e.g. 'liquefaction' (Castro and Poulos 1977; Ishihara et al. 1991), 'complete liquefaction' (Mohamad and Dobry 1986), 'flow-type failure' (Vaid and Chern 1983; Yang and Sze 2011; Yoshimine and Ishihara 1998) and 'cyclic instability' (Baki et al. 2012; Ishihara et al. 1975; Lo et al. 2010). However, the term "cyclic instability" is preferred in this study.

Cyclic instability can be observed for three different loading categories: (a) one-way cyclic loading when both loading and unloading phases are laid in one side of stress space (i.e. compression side or extension side); (b) two-way symmetrical loading when both loading and unloading phases are laid in both of stress space with equal magnitude; (c) two-way non-symmetrical loading when loading and unloading phases are laid in both sides of stress space with uneven magnitude. Different loading categories also produce different paths during cyclic loading. Many researchers suggested that static liquefaction and cyclic instability are linked and they have similar failure mechanism (Baki et al. 2012, 2014; Mohamad and Dobry 1986; Rahman et al. 2014). The static liquefaction trigger at instability zone i.e. at peak stress ratio state, $\eta_{IS} = q/p'$ (Mizanur and Lo 2012; Nguyen et al. 2017; Rabbi et al. 2018; Rahman et al. 2017; Zhang et al. 2018). Cyclic instability trigger when the cyclic stress path reaches the instability zone of an identical monotonic loading test. However, the evaluation of the link between cyclic instability and static liquefaction from constitutive modelling point of view is still limited.

There has been significant development of constitutive modelling within the framework of critical state soil mechanics (CSSM). It was found that both monotonic and cyclic liquefaction behaviours were closely related with a state parameter (ψ), which is defined as the difference of current e and e at critical state (CS) for the same p' by Been and Jefferies (1985). Then, ψ has been commonly used as the state index for sands and other granular materials (Nguyen and Rahman 2017; Nguyen et al. 2015; Rahman et al. 2018). The relationships between ψ and other parameters related to material behaviour (e.g. instability stress ratio, cyclic resistance ratio, etc.) have also been widely investigated in liquefaction study (Mizanur and Lo 2012; Nguyen et al. 2015, 2017; Rahman et al. 2008). A state-dependent constitutive model based on ψ, SANISAND (Dafalias and Manzari 2004; Li and Dafalias 2000), is adopted to calibrate the testing data in this study.

Therefore, the objective of this study is to adopt a constitutive model within the CSSM framework to develop constitutive relation for both monotonic and cyclic instability behaviour. The constitutive model was then used to predict static liquefaction behaviour to conform applicability of the model. The model was then extended to simulate cyclic instability behaviour to evaluate the link between static and cyclic instability behaviour.

2 Materials

The sand used in this study Sydney sand. It is a clean uniform size quartz sand (SP) with a mean size of 0.30 mm. The grain size distribution curve of Sydney sand is shown in Fig. 1. The detail properties of Sydney sand, including SEM photograph, can be found in earlier publication (Lo et al. 2010). The triaxial testing system and triaxial testing results of Sydney sand can be found in other publication (Bobei et al. 2009, 2013; Rahman and Lo 2014). A constitutive model for monotonic loading was developed earlier by Rahman et al. (2014), however this publication presents a preliminary development of a constitutive model that can predict both monotonic and cyclic loading behaviour of Sydney sand.

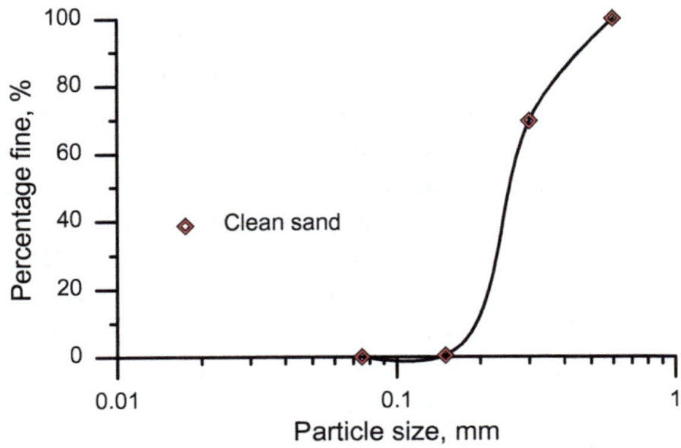

Fig. 1. Grain size distribution of Sydney sand.

3 Model Formulations and Calibrations

In triaxial condition, ε_1 and ε_3 are axial and lateral strains respectively, whereas σ'_1 and σ'_3 are major and minor principal effective stresses respectively. Other triaxial variables are deviatoric strain ($\varepsilon_q = 2/3[\varepsilon_1 - \varepsilon_3]$), volumetric strain ($\varepsilon_v = \varepsilon_1 + 2\varepsilon_3$), deviatoric stress ($q = \sigma'_1 - \sigma'_3$) and mean effective stress ($p' = [\sigma'_1 + 2\sigma'_3]/3$).

3.1 Model Yield Function and Incremental Formulations

For a more complete triaxial model addressing cyclic loading, the yield function is represented by a cone in multiaxial or a wedge in triaxial stress space, as shown in Manzari and Dafalias (1997), and Dafalias and Manzari (2004). The yield function can be expressed as:

$$f = |\eta - \alpha| - m = 0 \tag{1}$$

where η is stress ratio ($\eta = q/p'$), α and m are both stress ratio quantities. The model consists of a series of constitutive equations for elastic and plastic strain increment components. The additive decomposition of strain increment, as in one of SANISAND model (Li and Dafalias 2000), is often considered as:

$$d\varepsilon_q = d\varepsilon_q^e + d\varepsilon_q^p = \frac{dq}{3G} + \frac{d\eta}{H} \quad (2a)$$

$$d\varepsilon_v = d\varepsilon_v^e + d\varepsilon_v^p = \frac{dp'}{K_p} + d\left|\frac{d\eta}{H}\right| \quad (2b)$$

where the superscripts "*e*" and "*p*" denote elastic and plastic components, respectively; G and K_p are elastic shear and bulk moduli; H is the plastic hardening modulus; d is dilatancy. It should be noted that H is a variant of K_p, which is also the plastic hardening modulus from the simple triaxial model in Li and Dafalias (2000). In that model, K_p is also used to calculate the incremental plastic deviatoric strain as the following equation:

$$d\varepsilon_q^p = \frac{p' d\eta}{K_p} \quad (3)$$

Substituting Eq. (2a) into (3):

$$K_p = Hp' \quad (4)$$

3.2 Elastic Modulus

The elastic shear modulus (G) of a sandy soil can be expressed as a function of e and p' by the following equation (Goudarzy et al. 2016, 2017; Hardin and Richart 1963; Iwasaki and Tatsuoka 1977; Rahman et al. 2012, 2014):

$$G = C_g p_a \frac{(2.97 - e)^2}{1 + e} \left(\frac{p'}{p_a}\right)^{n_g} \quad (5)$$

where C_g and n_g are model parameters which are calibrated on the q-ε_q path at early stage of shearing; p_a is atmospheric pressure ($\approx$101 kPa). Note, instead of maximum elastic shear modulus (G_0) in common elasticity studies (Hardin and Richart 1963; Iwasaki and Tatsuoka 1977), C_g and n_g are adopted as elastic parameters in the monotonic loading model for Sydney sand by Rahman et al. (2014). The incremental elastic bulk modulus (K) for sands is given by:

$$K = \frac{2(1 + \nu)}{3(1 - 2\nu)} G \quad (6)$$

where ν is Poisson's ratio which is also calibrated on the q-ε_q path at small strains.

3.3 Plastic Modulus

Under drained loading condition, η rises, but is bounded by the peak stress ratio M^b, which is also called bounding stress ratio. Therefore, the plastic hardening modulus (H) is controlled by M^b and can be expressed as the following equation:

$$H = h(M^b - s\eta) \tag{7}$$

where s is loading direction (±1) and h is the function of the state variables:

$$h = b_0/|\eta - \eta_{in}| \tag{8}$$

where b_0 can be defined as:

$$b_0 = C_g h_0(1 - c_h e)(p'/p_a)^{-n_g} \tag{9}$$

It should be noted that the formulation of b_0 in this study is modified based on C_g and n_g, instead of G_0 in Dafalias and Manzari (2004). This modification might help to model the sandy materials having high fine contents e.g. Sydney sand with fines content (Rahman et al. 2014).

3.4 Dilatancy

Dilatancy, proposed by Rowe (1962), relates to the difference of current η from the dilatancy stress ratio (M^d). The definition of M^d is similar to the phase transformation line in Ishihara et al. (1975). The dilatancy equation in this model, hence, can be written as:

$$d = A_d(M^d - s\eta) \tag{10}$$

where A_d is the function of the state and can be expressed as:

$$A_d = A_0(1 + \langle sz \rangle) \tag{11}$$

where A_0 is model parameter; z is fabric-dilatancy internal variable; the MacCauley brackets $\langle \ldots \rangle$ means that $\langle sz \rangle = sz$ if $sz > 0$ and $\langle sz \rangle = 0$ if $sz \leq 0$. According to Dafalias and Manzari (2004), the change of fabric in this model influences the dilatancy behaviour. The incremental fabric change can be defined as

$$dz = -c_z\langle -d\varepsilon_v^p \rangle (sz_{max} + z) \tag{12}$$

where c_z and z_{max} are both model parameters.

3.5 Critical State Behaviour

According to the CSSM framework, a set of CS data points can form a unique CSL in the classical e-log(p') space and the critical state stress ratio (M) in the q-p' space is also unique for a soil (Schofield and Wroth 1968). The CSL in the e-log(p') and the q-p' space can be expressed by the Eqs. 13a and 13b respectively.

$$e_{CS} = e_{lim} - \Lambda(p'/p_a)^{\xi} \tag{13a}$$

$$M = q_{CS}/p'_{CS} \tag{13b}$$

where e_{lim}, Λ and ξ are CS parameters that present the y-intercept, slope and curvature of CSL respectively; and M is the critical state stress ratio. It should be noted that the CS parameters for Sydney sand are determined by fitting all CS data points in the q-p' and e-log(p') spaces. The detail can be found in Rahman et al. (2014). The CSLs for Sydney sand in both q-p' and e-log(p') spaces are shown in Fig. 2. The state parameter (ψ) proposed by Been and Jefferies (1985) can be defined as:

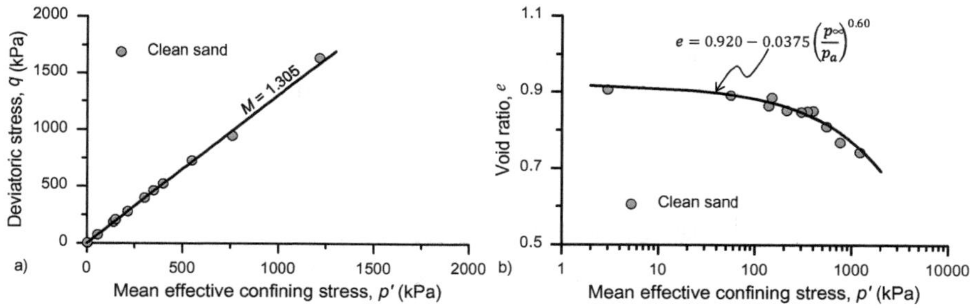

Fig. 2. CSL for Sydney sand: (a) in q-p' and (b) in e-log(p') spaces.

$$\psi = e_{CS} - e \tag{14}$$

All stress ratios, M, M^d and M^b, are considered as material parameters, as they represents the material states for a soil. Under drained condition, a dense specimen initially exhibits volume contraction and phase-transforms to volume dilation. The transition point is called characteristic state (Lade and Ibsen 1997; Nguyen et al. 2018), which is controlled by M^d in this model. The characteristic state often occurs at early stage of shearing. After the characteristic state, a drained peak stress is attained, which is controlled by M^b in this model. After the initial peak, strain softening takes place until reaching the CS or M-line. In theory, $M^b > M > M^d$. The stress ratios, M^d and M^b, can be defined as:

$$M^b = Me^{-n^b\psi} \tag{15a}$$

$$M^d = Me^{n^d\psi} \tag{15b}$$

where n^b is hardening parameter which is calibrated at the drained peak stress, whereas n^d is dilatancy parameter which is calibrated at the characteristic state. After determining n^d, the other dilatancy parameter (A_0) in Eq. (11) can be calibrated on the ε_v-ε_q path in monotonic drained loading. All model parameters in this study is listed in Table 1.

Table 1. Model parameters

Parameter	Variable	Value
Elastic parameter	C_g	186
	n_g	0.75
	ν	0.25
Critical state parameter	M	1.305
	$e_{\lim}$	0.920
	Λ	0.0375
	ξ	0.60
Yield surface	m	0.01
Plastic parameter	h_0	20
	c_h	0.90
	n^b	1.06
Dilatancy parameter	A_0	1.06
	n^d	0.5
Fabric parameter	z_{max}	6
	c_z	600

4 Results and Model Validations

The monotonic test (S-MII-00-10, e_0 = 0.880 and p'_0 = 850 kPa), which shows static liquefaction, is considered here to compare with the model prediction as shown in Fig. 3. Since the model is for cyclic loading, a monotonic path means only a loading cycle in compression side that produce a bounding surface as monotonic loading path in compression side. The predicted effective stress path by the model is presented in the q-p' space (Fig. 3a). While slight variation observed between test results and model prediction, it is important that the characteristic features if undrained behaviour matched very well e.g. critical state, instability etc. As mentioned earlier, the static liquefaction triggers at the instability zone and in this case the instability zone, as shown by a shaded wedge in Fig. 3a, was identical with the experimental data. Figure 3c shows the development of pore water pressure (Δu) during shearing. In both test and model prediction, it is evident that Δu is built up until reaching its maxima at CS or liquefaction.

Many literature reported that cyclic instability behaviour is linked with static liquefaction behaviour (Baki et al. 2012); cyclic instability occurs when the effective stress path in the q-p' space reaches the instability zone (or instability stress ratio) from a corresponding monotonic loading test. To evaluate whether the cyclic model has the capability to capture this feature, the model simulations for cyclic loading condition

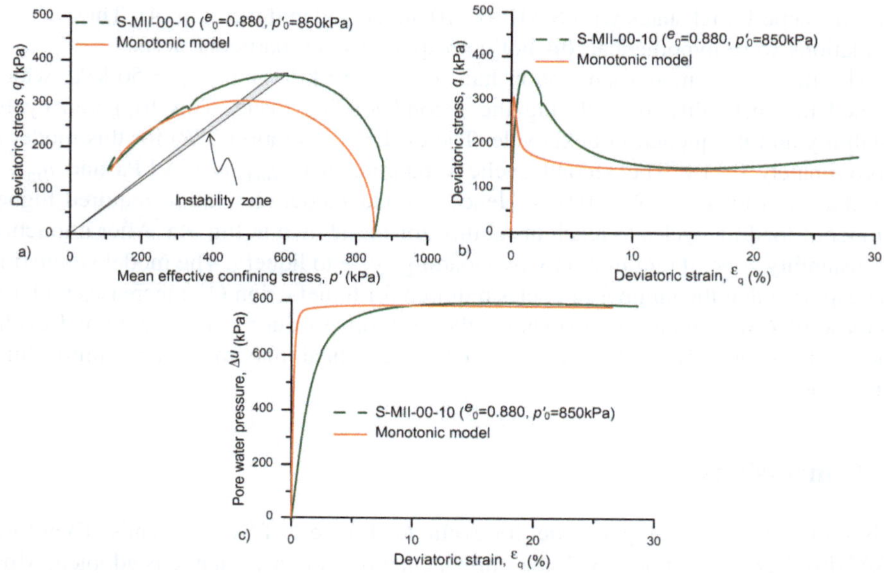

Fig. 3. Model prediction for monotonic loading test (S-MII-00-10): (a) in q-p' and (b) in q-ε_q and (c) in Δu-ε_q spaces.

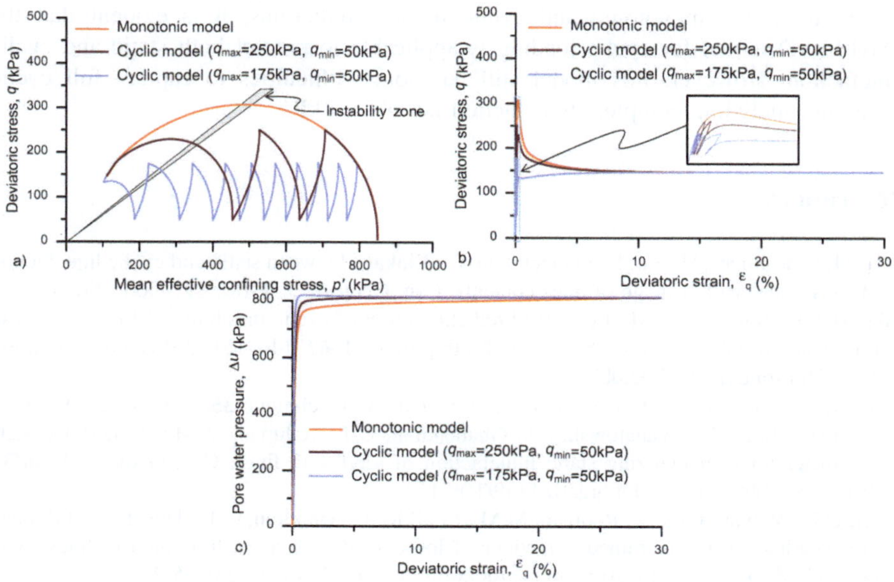

Fig. 4. Model prediction for cyclic loading test (S-MII-00-10): (a) in q-p' and (b) in q-ε_q and (c) in Δu-ε_q spaces.

with the same initial states with S-MII-00-10 are considered (see Fig. 4). These model simulations were hypothetical, dis not correspond to any particular tests.

The first cyclic model simulation had q_{max} = 250 kPa and q_{min} = 50 kPa, which reached the instability zone during the second loading cycle. This triggered cyclic instability and it liquefied in next cycle. The cyclic stress ratio (*CSR*) for this model is approximately 0.118. The second cyclic simulation had q_{max} = 175 kPa and q_{min} = 50 kPa i.e. a lower *CSR* of 0.074. Hence, second model simulation required higher number of loading cycles to reach instability zone as shown in Fig. 4a. After it touched the instability zone, it required only two loading cycle to liquefy. The model simulation also captured that the number of cycles required for liquefaction (N_L) increases with the decrease of *CSR*. Figure 4c also shows the built-up Δu during monotonic and cyclic loading tests. All Δu paths clearly reach an identical maxima when approaching liquefaction.

5 Conclusions

This study aims to develop a model for cyclic loading test of Sydney sands. Therefore, a SANISAND cyclic model with the implementation of fabric change is adopted. Most of model parameters can be obtained from the monotonic loading tests, which were previously reported for a monotonic model of Sydney sand. It is evident that the cyclic instability triggers when it touches the instability zone obtained from a monotonic loading test. However, further investigation requires to justify this in detail.

Based on the monotonic and cyclic model predictions, it is evident that the SANISAND model for cyclic loading is applicable to predict both static and cyclic liquefaction. However, this model still has some difficulties to capture full cyclic behaviour, including complete liquefaction i.e. $p' \approx 0$ kPa.

References

Baki, M.A., Rahman, M., Lo, S., Gnanendran, C.: Linkage between static and cyclic liquefaction of loose sand with a range of fines contents. Can. Geotech. J. **49**(8), 891–906 (2012)

Baki, M.A.L., Rahman, M.M., Lo, S.R.: Predicting onset of cyclic instability of loose sand with fines using instability curves. Soil. Dyn. Earthq. Eng. **61–62**, 140–151 (2014). https://doi.org/10.1016/j.soildyn.2014.02.007

Been, K., Jefferies, M.G.: A state parameter for sands. Geotechnique **35**(2), 99–112 (1985)

Bobei, D.C., Lo, S.R., Wanatowski, D., Gnanendran, C.T., Rahman, M.M.: A modified state parameter for characterizing static liquefaction of sand with fines. Can. Geotech. J. **46**(3), 281–295 (2009). https://doi.org/10.1139/T08-122

Bobei, D.C., Wanatowski, D., Rahman, M.M., Lo, S.R., Gnanendran, C.T.: The effect of drained pre-shearing on the undrained behaviour of loose sand with a small amount of fines. Acta Geotech. **8**(3), 311–322 (2013). https://doi.org/10.1007/s11440-012-0195-2

Castro, G., Poulos, S.J.: Factors affecting liquefaction and cyclic mobility. J. Geotech. Eng. Div. **103**(6), 501–506 (1977)

Dafalias, Y.F., Manzari, M.T.: Simple Plasticity Sand Model Accounting for Fabric Change Effects. J. Eng. Mech. **130**(6), 622–634 (2004). https://doi.org/10.1061/(ASCE)0733-9399(2004)130:6(622)

Goudarzy, M., Rahemi, N., Rahman, M.M., Schanz, T.: Predicting the Maximum Shear Modulus of Sands Containing Nonplastic Fines. J. Geotech. Geoenvironmental Eng. **143**(9), 06017013 (2017). https://doi.org/10.1061/(ASCE)GT.1943-5606.0001760

Goudarzy, M., Rahman, M.M., König, D., Schanz, T.: Influence of non-plastic fines content on maximum shear modulus of granular materials. Soils Found. **56**(6), 973–983 (2016). https://doi.org/10.1016/j.sandf.2016.11.003

Hardin, B.O., Richart, F.E.J.: Elastic wave velocities in granular soils. J. Soil Mech. Found. Div. ASCE **89**(1), 33–65 (1963)

Ishihara, K., Tatsuoka, F., Yasuda, S.: Undrained deformation and liquefaction of sand under cyclic stresses. Soils Found. **15**(1), 29–44 (1975)

Ishihara, K., Verdugo, R., Acacio, A.A.: Characterization of cyclic behaviour of sand and post-seismic stability analysis. In: Proceedings of 9th Asian Regional Conference on soil mechanics and foundation engineering, 45–68 (1991)

Iwasaki, T., Tatsuoka, F.: Effects of grain size and grading on dynamic shear moduli of sands. Soils Found. **17**(3), 19–35 (1977)

Lade, P.: Instability and liquefaction of granular materials. Comput. Geotech. **16**(2), 123–151 (1994)

Lade, P.V., Ibsen, L.B.: A study of the phase transformation and the characteristic lines of sand behaviour. In: Proceedings of International Symposium on Deformation and Progressive Failure in Geomechanics, Nagoya, pp. 353–359 (1997)

Li, X., Dafalias, Y.: Dilatancy for cohesionless soils. Geotechnique **50**(4), 449–460 (2000)

Lo, S., Rahman, M.M., Bobei, D.: Limited flow characteristics of sand with fines under cyclic loading. Geomech. Geoengin. Int. J. **5**(1), 15–25 (2010)

Manzari, M.T., Dafalias, Y.F.: A critical state two-surface plasticity model for sands. Geotechnique **47**(2), 255–272 (1997)

Mizanur, R.M., Lo, S.: Predicting the onset of static liquefaction of loose sand with fines. J. Geotech. Geoenvironmental Eng. **138**(8), 1037–1041 (2012)

Mohamad, R., Dobry, R.: Undrained monotonic and cyclic triaxial strength of sand. J. Geotech. Eng. **112**(10), 941–958 (1986)

Nguyen, H.B.K., Rahman, M.M.: The role of micro-mechanics on the consolidation history of granular materials. Aust. Geomech. **52**(3), 27–36 (2017)

Nguyen, H.B.K., Rahman, M.M., Cameron, D.A.: Undrained behavior of sand by DEM study. In: International Foundations Congress Equipment Expo (IFCEE 2015), ASCE, pp. 182–191 (2015a). https://doi.org/10.1061/9780784479087.019

Nguyen, H.B.K., Rahman, M.M., Cameron, D.A., Fourie, A.B. The effect of consolidation path on undrained behaviour of sand - a DEM approach. In: Computer Methods and Recent Advances in Geomechanics, CRC Press, pp. 175–180 (2015b). https://doi.org/10.1201/b17435-27

Nguyen, H.B.K., Rahman, M.M., Fourie, A.B.: Undrained behaviour of granular material and the role of fabric in isotropic and K0 consolidations: DEM approach. Géotechnique **67**(2), 153–167 (2017). https://doi.org/10.1680/jgeot.15.P.234

Nguyen, H.B.K., Rahman, M.M., Fourie, A.B.: Characteristic behaviour of drained and undrained triaxial tests: a DEM study. J. Geotech. Geoenvironmental Eng., In Press (2018) https://doi.org/10.1061/(asce)gt.1943-5606.0001940

Rabbi, A.T.M.Z., Rahman, M.M., Cameron, D.A.: Undrained Behavior of Silty Sand and the Role of Isotropic and K0 Consolidation. J. Geotech. Geoenvironmental Eng. **144**(4), 04018014 (2018). https://doi.org/10.1061/(ASCE)GT.1943-5606.0001859

Rahman, M., Baki, M., Lo, S.: Prediction of undrained monotonic and cyclic liquefaction behavior of sand with fines based on the equivalent granular state parameter. Int. J. Geomech. **14**(2), 254–266 (2014a). https://doi.org/10.1061/(ASCE)GM.1943-5622.0000316

Rahman, M., Lo, S.-C., Dafalias, Y.: Modelling the static liquefaction of sand with low-plasticity fines. Géotechnique **64**(11), 881–894 (2014b)

Rahman, M.M., Cubrinovski, M., Lo, S.R.: Initial shear modulus of sandy soils and equivalent granular void ratio. Geomech. Geoengin. **7**(3), 219–226 (2012). https://doi.org/10.1080/17486025.2011.616935

Rahman, M.M., Lo, S., Gnanendran, C.: On equivalent granular void ratio and steady state behaviour of loose sand with fines. Can. Geotech. J. **45**(10), 1439–1456 (2008)

Rahman, M.M., Lo, S.R.: Undrained behaviour of sand-fines mixtures and their state parameters. J. Geotech. Geoenvironmental Eng. **140**(7), 04014036 (2014). https://doi.org/10.1061/(ASCE)GT.1943-5606.0001115

Rahman, M.M., Nguyen, H., Rabbi, Z.: Undrained behaviour of sand under isotropic and K0-consolidated condition: experimental and DEM approach. In: The 19th International Conference on Soil Mechanics and Geotechnical Engineering, pp. 493–496 (2017)

Rahman, M.M., Nguyen, H.B.K., Rabbi, A.T.M.Z.: The effect of consolidation on undrained behaviour of granular materials: a comparative study between experiment and DEM simulation. Geotech. Res., In press (2018)

Rowe, P.W.: The stress-dilatancy relation for static equilibrium of an assembly of particles in contact. Proc. R. Soc. Lond. A **269**(1339), 500–527 (1962)

Schofield, A.N., Wroth, P.: Critical State Soil Mechanics, McGraw-Hill (1968)

Vaid, Y., Stedman, J., Sivathayalan, S.: Confining stress and static shear effects in cyclic liquefaction. Can. Geotech. J. **38**(3), 580–591 (2001)

Vaid, Y.P., Chern, J.C.: Effect of static shear on resistance to liquefaction. Soils Found. **23**(1), 47–60 (1983)

Yamamuro, J.A., Covert, K.M.: Monotonic and cyclic liquefaction of very loose sands with high silt content. J. Geotech. Geoenvironmental Eng. **127**(4), 314–324 (2001)

Yang, J., Sze, H.: Cyclic behaviour and resistance of saturated sand under non-symmetrical loading conditions. Géotechnique **61**(1), 59–73 (2011)

Yoshimine, M., Ishihara, K.: Flow potential of sand during liquefaction. Soils Found. **38**(3), 189–198 (1998)

Zhang, J., Lo, S.-C.R., Rahman, M.M., Yan, J.: Characterizing Monotonic Behavior of Pond Ash within Critical State Approach. J. Geotech. Geoenvironmental Eng. **144**(1), 04017100 (2018). https://doi.org/10.1061/(ASCE)GT.1943-5606.0001798

Stability and Reliability Analysis of the Slope of Haiqar Dam Supports Using Key Groups Method

Mehdi Mokhberi(✉) and Bahman Jahan Bekam Fard

Department of Geotechnical Engineering, Islamic Azad University, Estahban Branch, Estahban, Iran
m_mokhberi@iau.ac.ir

Abstract. The study of engineering and mechanical behavior of rocks began after the collapse of jointed rocks of Vajont Dam in October 1963 in which more than 2000 people were killed. Since the dam was not damaged in this event, the overlap of engineering geology and geotechnical engineering is highly significant. Therefore, the analysis of the stability and reliability of the slope of dam stone supports, especially that of jointed stones, is one of the most important issues considered in the design of dams which is accessible through the implementation of a geotechnical model. In spite of all of the scientific developments in the recent decades, this field still faces complicated problems due to the great diversity of factors affecting dam behaviors especially geotechnical conditions. Moreover, ignoring these problems can lead to project failure and great financial and human damages. One of the techniques of stability analysis is to find appropriate 3D geometrical models for rock masses including a set of simulated geotechnical blocks of masses which are formed based on topography position, the model limit, tectonic blocks, and the existing discontinuities using definite, statistical, or geotechnical methods. The findings of the present study can have significant implications in geotechnical studies for the estimation of blocks volumes and the analysis and mechanical modeling on blocks, especially the walls loss modeling and their stability analysis. Accordingly, Haiqer Reservoir Dam, located in Fars province in Iran, was analyzed in the present study. The results indicated that the left support has a low reliability and therefore, its loss risk is very high. In contrast, the left support has better stability conditions with less probable great loss risks. It was also noted that if the blocks are taken into account as a whole they can form a key group which is potentially more dangerous than a single key block.

Keywords: Rock mechanics · Dam abutment · Slope stability
Key groups · Block theory

1 Introduction

In rock mechanics category, the Key Groups method can be employed for geotechnical engineering projects and the Key Block Analysis method can be used on geometric models with block sets. Results of limit equilibrium analysis on some blocks show that

H. Shehata and C. S. Desai (Eds.): GeoMEast 2018, SUCI, pp. 87–100, 2019.
https://doi.org/10.1007/978-3-030-01926-6_6

if such blocks are not maintained they may become destabilized and such instability progressively spills over other blocks. Such blocks are called *key blocks*. The results of research suggest that the stability analysis of rock masses with fissure can lead to the identification of block groups, not only a single block so that the blocks taken together can form a key group which is potentially more dangerous than a single key block.

There are two main methods for the analysis of key blocks: (1) Vector approach (Warburton 1981) and (2) Graphical approach (Goodman-Sh, 1995). Both methods only consider sliding movement of a block and they do not analyze the rotational movement of a given block. Other researchers e.g. Bray and Goodman (1981), Fulvio and Tonon (2000) and Lin et al. (1987) have expanded vector analysis of rotational movements and a few researchers have tried to correct key blocks approach. For instance, Wibowo (1997) has studied the Second Order Key Blocks Search using key blocks analysis based on Goodman-Shi's method. Another approach proposed to the generalized key blocks method is a grouping technique to analyze all blocks neighboring a key block. This technique, called Key Groups Method (KGM), searches a key group that is more unstable than single key blocks performed using a progressive stability analysis (Yarahmadi-Verdel, 2003). An extension of KGM based on Sarma method, abbreviately called SKGM, has been proposed by Yarahmadi-Verdel (2004) to analyze the rock slope stability.

Geotechnical models of rock masses include a set of geotechnical simulated blocks of masses formed statistically based on the topographical position, the model limit, tectonic blocks, and existing discontinuities.

The mechanical model of a rock mass aims to analyze the mechanical stability (either static or dynamic) of a given rock mass and includes the analysis of geomechanical parameters of the mass, situ stress conditions, groundwater conditions, rules governing the behavior and the strength of the rock mass (resistance indices of fractures and mechanical behaviors of the environment), methods of stability analysis, and finally making inferences and conclusions about the mechanical stability (such as the confidence coefficient or the confidence interval). In the stability analysis of slopes based on a statistical method we cannot simply satisfied to a number such as confidence interval expressed definitely or indefinitely as rocky and dusty environments are often highly heterogeneous and anisotropic. Accordingly, the use of probabilistic methods to apply more accurate parameters affecting stability seems necessary.

A probabilistic model was developed to identify and evaluate ways to reduce the ambiguity of the two above mentioned models. The results of the model included event probability, reliability or the level of geotechnical risk that are used to study statistical characteristics of effective parameters and variables in the models, methods of probabilistic analysis (e.g. reliability and simulation), and ultimately probabilistic deduction and conclusions.

2 Materials and Methods

2.1 Topographical Features of Haiqer Dam

Haiqer Reservoir Dam is situated at a place called Haiqer Strait on Firuz Abad River about 45 km southeast of Firuz Abad City, Fars Province. This roller compacted concrete (RCC) dam was built to control untimely floods flowing into the dam site. The slope of the valley floor on both sides of the dam in the upstream is 160 m ahead of the dam axis and in the downstream it is 240 m far away from the dam axis that changes along the river. The slopes of the river are 62% and 8%, respectively. The width of valley at the valley floor at the axis site is about 37.5 m. Besides, the width at levels of 1130, 1160, and 1180 m is114, 163 and 194.5 m, respectively. The slope at the right abutment in the upper parts is 65° and it increases when moving downward; reaching to 41°. The slope at the right abutment in the parts above the dam crest is less than other parts and sometimes it is 8°. However, the slope increases at the abutment and reaches to 62° but it decreases in areas near the floor; reaching to18°. The valley axle at the dam trunk is directed north-eastward and south-westward at a 60° angle. At a distance of 160 m far away from the trunk, the axle is 16°. Finally, at a distance of 85 m from the dam trunk; the valley axle changes and is placed in a northern-southern direction.

2.2 Stability Analysis and Reliability of Slope of Walls and Abutments of Haiqer Dam

2.2.1 Geometric-Geotechnical Modeling of Abutments of Haiqer Dam

Discontinuities have a profound effect on the mechanical and hydraulic behavior of rocks. They play a vital role in determining the size and the shape of stone blocks in the rock masses under study and, finally, in the creation of a geometrical model for the rock masses. To do so, first a topographic model of the masses and the area under study will be produced. The rock masses will be divided into statistically homogeneous regions based on structural features and using the main structures such as faults. Accordingly, in order to know these regions; joint study operations including logging, plotting, and analysis were performed as well as statistical analysis on geometric information. Finally, the geometrical modeling of rock masses was prepared based on information from the above steps and the use of different methods of discontinuity network simulation (Fig. 1).

The first step of geometric modeling is having knowledge of the area under study and its topography. Therefore, the boundaries of the model and sections under analysis were determined using the topography of the area (CAD files and maps plotted by Energy and Water Engineers Corporation). Besides, the logs of exploratory boreholes of the position of Haiqer Dam can provide some general ideas of the model of fractures of the area, details of joints and fractures as well as the morphological and lithological studies (See Fig. 2).

2.2.2 Determining Statistically Homogeneous Areas (The Structure)

To reduce simulation variations and errors in statistical analysis, the division of population into smaller communities is not done by taking into account the amount of

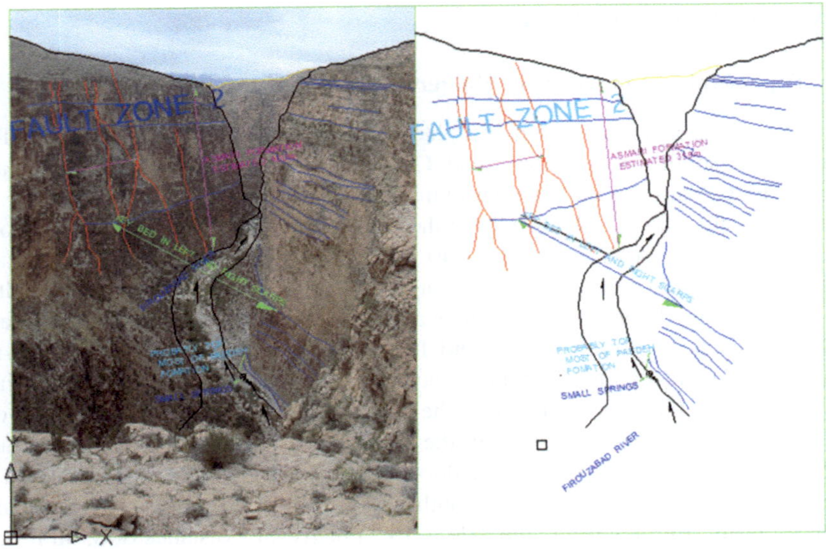

Fig. 1. A view of Haiqer Valley and Firoozabad River Floor. Look at nearly horizontal layers and fault zone No. 2 where there are many karstic channels to excavate the dam water so that, based on piezometric speculations, the groundwater is at present fleeting from the dam area. The first area of the fleeting water is at the valley bottom at the right side of the above picture seen in the form of small springs (shown by arrows).

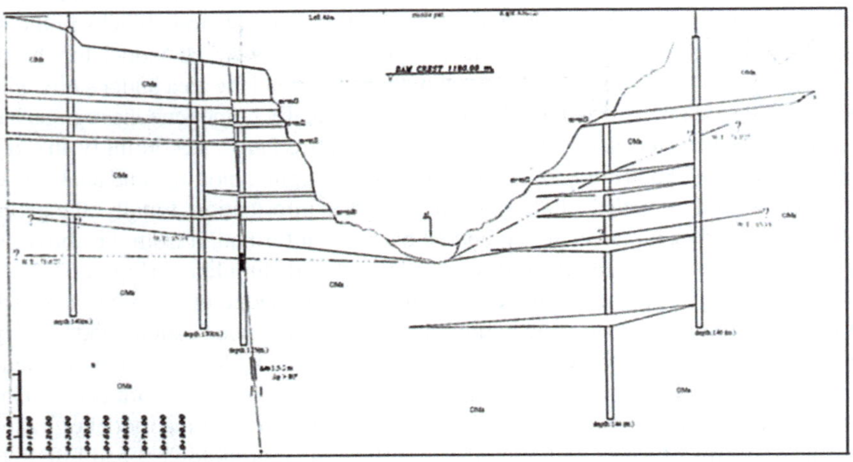

Fig. 2. Cross-section under study

reliable information. It is also necessary when simulating discontinuities. Therefore, the rock mass in the area under study was analyzed to divide it into blocks with similar characteristics of geological structures. First, the mass was divided into a single tectonic block due to the absence of Fault F1. The assumption used to divide statistically homogenous regions is usually based on the main assumptive structures. Otherwise, the areas of the same statistical moments are combined again.

2.2.3 Structural Factors

The dominant structural elements after the main faults in jointed rock masses are discontinuities (joints, layering, schistosity, cross-slope, etc.) whose features in statistically homogeneous regions are focused upon in geometric modeling. Joint study is one of the most important stages of geometric modeling that happens through the following steps:

- Logging
- Plotting and Classification
- Statistical Analysis

2.2.4 Joints Logging

Joint logging operations are done by surface projects, normally by jogging lines parallel to the front of activities done by each tectonic block. However, the geometrical characteristics of discontinuities in the region under study were obtained through the data from the boreholes.

2.2.5 Platting and Classification of Fractures

The data from the analysis of loggings of the existing fractures in the region as well as information mined from the classified layers in the region were used to simulate and model the fractures separately using statistical techniques by Dips Software Package.

A total number of 460 joints were analyzed in the region under study. The drawing of flow charts based on Schmidt's Distribution Function, the 2% floating circle, equal levels network by Dips Software shows three prominent categories of joints. Low concentrated joints were also visible that were regarded as layered surfaces.

The windows submenu in this software makes it possible to separate and classify the joints and to store them in separate files. Accordingly, the interval differences between the longitudinal loggings give way to the possibility of measuring the horizontal distances between joints. Eventually, statistical distribution functions of each component of discontinuities including slope, slope direction, and spacing using Dips and Minitab software were discussed in the present study.

2.3 Statistical Analysis of Fractures

Geometrical modeling using statistical methods is one of the requirements for the distribution functions and statistical moments to determine features of each set of joints. This section discusses the slope profile, slope direction, and spacing of each set of joints. Using Minitab Software, the histograms for slope parameters and slope directions as well as the nature of their distribution function were analyzed and data normality was examined by the use of Anderson-Darling Test. If the p-value obtained from the test is greater than 0.005, it is assumed that the data have been distributed normally. Slope parameters and slope directions were regarded as normal in the present study. Figure 3 shows the histograms of the distribution of slope parameters and slope directions Joint Set 1:

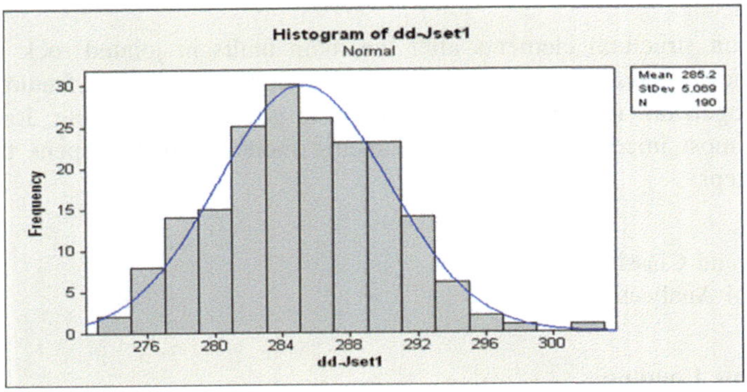

Fig. 3. Histogram of Slope Direction for Joint Set 1

The construction of 3D geometric models and the preparation of 2D cross-sections were performed by the use of Mathematica (3DGM) Software. The outputs of the software are presented in the form of 3D pages and the traces of discontinuities on cross-sections are shown as 2D pages. Theses outputs are used for numerical modeling and equilibrium point modeling (Fig. 4).

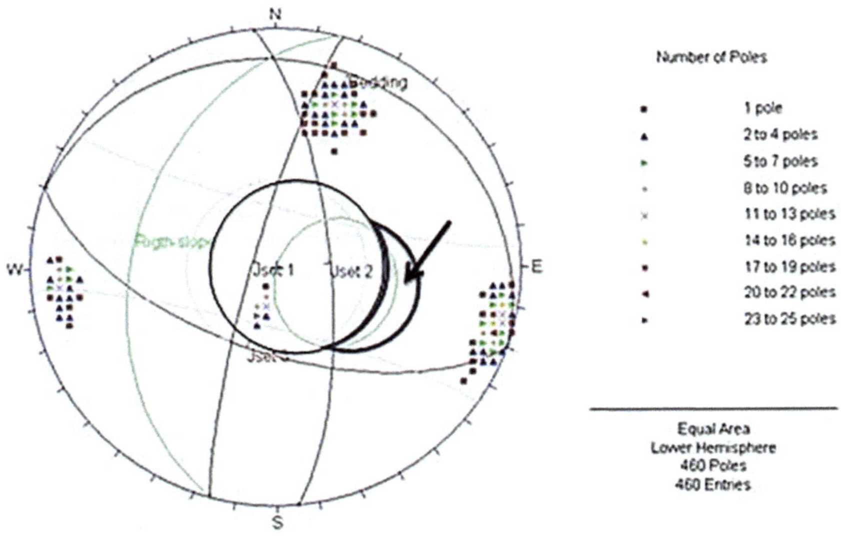

Fig. 4. The position of joint sets, wall facies, and polar points related to each discontinuity at the right abutment

Geological structures especially joint sets (discontinuities) can be simulated three dimensionally in Mathematica setting using geometrical-geotechnical modeling programs and statistical methods. The constructed 3D model provides detailed information about joints such as joint locations and geometrical structure of the formed blocks (corners, edges, and surfaces). Furthermore, database outputs are appropriate for the calculation of block sizes and the analysis of mechanical modeling on the blocks that are highly significant in geotechnical studies of dam abutments.

Given that the main cause of the loss of stone gables is structural reasons, in other words, joints play an important role in the instability, it can be said that the behavior affecting the rock masses is the same as the behavior of discontinuities. In the present study, however, due to the high fragmentation of the rock mass in the left section and the low value of Rock Mechanical Ranking (RMR), the rock mass under study was considered as a continuous mass and the main fault studied here is Fault F1. On the other hand, the main cause of instability in the form of potential discontinuities in the right section is the existence of joints. Therefore, the selection criteria for the stability analysis are the behavioral norms of the joints in question. As in situ stresses are negligent at this stage, the behavior of rock material at the right section is considered as rigid and non-transformable.

Mohr-Coulomb Index was used as behavioral criterion to measure strength properties and behavioral analysis. This index was employed in KGM by UDEC Software after determining the strength properties.

To analyze the behavior of the stone walls of the area under study, joint behavioral indices were used for the right part of the walls and Hook Brown Behavioral Index was used for the left section of the walls. In addition, it was attempted to joint strength parameters using laboratory tests and in situ experiments. Accordingly, the obtained values from the Rock Mechanical Ranking (RMR) were used to measure mechanical parameters of joints including normal and cutting hardness of joint surfaces that were employed for numerical modeling.

2.4 Two-Dimensional Stability and Reliability Analysis of River Walls

The stability and the strength of walls were analyzed by the use of physical and mechanical properties of the masses and relationships governing the strength and the behavior of walls.

The proposed method of analysis used for the 2D mechanical modeling was Key Groups Methods (Yarahmadi-Verdel 2004) which is based on the equilibrium point analysis. Finally, the results of this method were compared with the results from numerical analysis of the discrete elements method by UDEC Software.

After determining the geometric properties of the joints by statistical methods, 3D geometric modeling was performed statistically using a model developed in Mathematica programming environment to provide a two-dimensional analysis of radial cross-section radius at the place where the river has a curvature. Furthermore, the mechanical model of the sections was implemented on geometric modeling and the analysis was performed statically (point equilibrium) using the key groups method.

The probabilistic model of the walls was also constructed to evaluate and reduce the ambiguities existing in the mechanical model. Variables included in this model were:

first and second order statistical moments (mean and SD), strength parameters of joint surfaces (c and φ), density of rock material (γ), and pairwise covariance values of the components. The information from the statistical data of the above parameters was used for subsequent analysis.

In probabilistic modeling, FORM method was used for the performance of reliability analysis. At this stage, the reliability analysis was performed based on the reliability coefficient and the possible loss of the 2D radial cross-section.

The output from each of the probabilistic analyses in accordance with stability indices presented in Tables 2 and 3 include a safety factor, the probability of a safety factor value less than one, Pr (SF < 1.0), the probability of a safety factor value less than one and a half, Pr (SF < 1. 5), and reliability index (β) for these two probabilities. After this stage, unstable groups and the most unstable key group were determined using the above stability indexes.

3 Stability Analysis of the Wall Using Key Groups Method

3.1 Stability Analysis of the Right Wall

As was mentioned earlier in the previous section, the rock mass is in the appropriate conditions based on the RMR value (e.g. 64–76). As a result, weak surfaces of the wall play a significant role in the stability or instability of the walls. Accordingly, models for potential loss arising from the discontinuity of the cross-section were studied using stereographic pictures by Dips Software.

The direction of the cross-section was determined based on the slope and the direction of the set of logged joints in a given block as was done for the block size. The set of the existing joints along with the position and the direction of the slope in cross-section of the right wall are shown on the stereographic network.

This cross-section is located in the direction of the dominant slope of the right wall. The general slope of the wall is 41° with a direction of 285°. Four major discontinuities can be seen in this block. The curved lines in the figure above at the two sides of the slope plate (marked with arrows) show the frontier of slope direction ± 20° proportional to the slope direction of the surface of the final wall. According to Goodman, the direction of the slope of each of these discontinuities or their overlap is the precondition for any possible initial instability. Given the slope and its direction at the final wall and the set of joints in this block, the loss resulting from such joints is not likely to happen. As shown in Fig. 5, the area formed by Daylight Envelope and friction angle circle represents the boundary of the planar fracture. Therefore, each pole placed in this area shows a plate that is likely to slide. None of poles of the joints and layered surfaces is located within Daylight Envelope and therefore the planar fracture is not likely to occur.

The overlap of the existing joints in sets 1 and 2 as well as layered surface and joint set 3 will create a wedge-like structure. The angle between the intersection line of the wedge formed by the joint sets 1 and 2 (marked with large arrows) with slope direction of the final wall is about 90° so it can be said that there is no risk of wedge instability in this part of the wall.

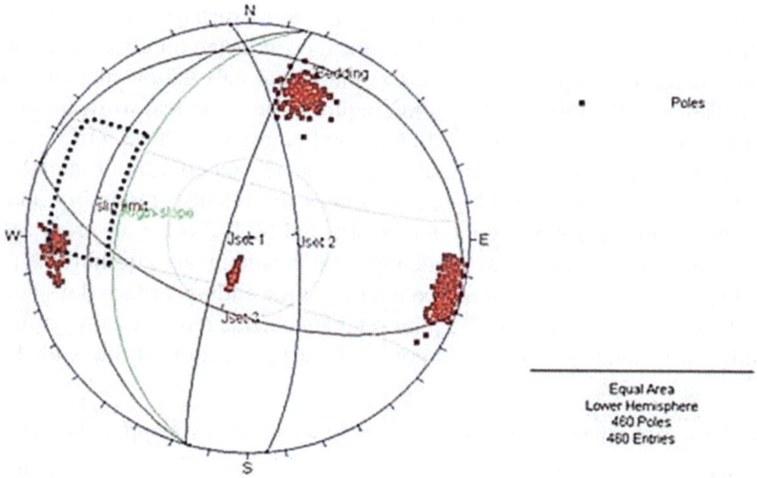

Fig. 5. The position of joint sets, wall facies, and the area with the possibility of rotational loss at the cross-section of the right wall

The angle between the intersection line of the wedge formed by the joint set 3 and layered surfaces (marked with small arrows) is approximately in the same direction of the slope of the final wall but with a slope directed horizontally. Therefore, given a friction angel of 30 to 35° for discontinuity surfaces and the corresponding friction cones, there is no risk of wedge loss in this area.

The joints from the joint set 2 can create rotational loss in the general slope of this cross-section (Fig. 5). Areas with a risk of rotational loss are shown by discontinued lines on the basis of the type of rotational loss with a friction angle of 30° for discontinuities at the final slope (Table 1).

Table 1. Parameters used for analysis and calculation methods for the right cross- section

Parameter	Calculation method	Unit	Mean	SD	Distribution
RMR	(Bieniawiski1989)		70	–	–
Equivalent elasticity modulus	$E_s = 2RMR - 100$ (Bieniawiski1989)	MPa	40	–	–
Joint normal harness	JKn = 10Es (Rode 1991)	GPam -1	400	–	–
Joint cutting harness	$10JKs \leq JKn \leq 100JKs$ (Rode 1991)	GPam -1	50	–	–
Volumetric weight of rock	Experimental	Kg/	2350	13	Normal
Joint adhesion	Barton-Bandis roperties	MPa	0.7	0.15	Normal
Joint friction angle	Barton-Bandis Index to	Degree	30	5.4	Normal

The strength parameters in this block and the joints surface was determined using the data from the joint logging and the experiments conducted on the block. Finally, the components used in the static analysis and numerical calculation methods are summarized in Table 2. Additional information required for numerical software like UDEC such as the joint cutting hardness ((JKN), joint normal hardness (JKS), and mass elasticity modulus are shown in the table. Also, the second column of the table shows the way the data are calculated. Figure 6b shows the results of stability analysis by FORM method and the use of the model constructed by KGM for the cross-section. In addition, the results of numerical modeling after reaching the equilibrium point are shown in Fig. 6a. Of course, it should be noted that the sliding surface does not exist in reality so the trace of layering and the joint set 3 are considered vertically on the section under study. In general, planar loss is not likely at this section of the wall (Fig. 7).

Table 2. Results of reliability analysis by KGM for unstable key group in the right section

Cycle	Method of analysis	Safety factor (SF)	P (SF < 1.0)	P (SF < 1.5)	Reliability index (β) (SF < 1.0)	Stability	Weight of unstable group (Ton)
444	FORM	1.95054	9.261×10^{-6}	0.02762	4.282	Stable slope	–

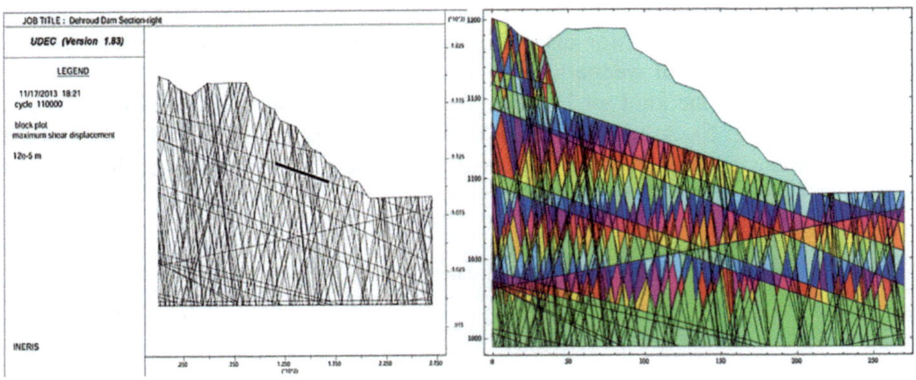

(a) (b)

Fig. 6. Comparison of analysis of the right cross-section by a) Numerical Method of Distinct Elements by the use of UDEC and b) Probabilistic Key Groups Method

The numerical model shows the shear surface based on the surface of the most unstable key group where the maximum sliding value is 1.2×10^{-4}.

The results of the reliability analysis using Key Groups Method has been shown in the following table for the group with the lowest safety factor.

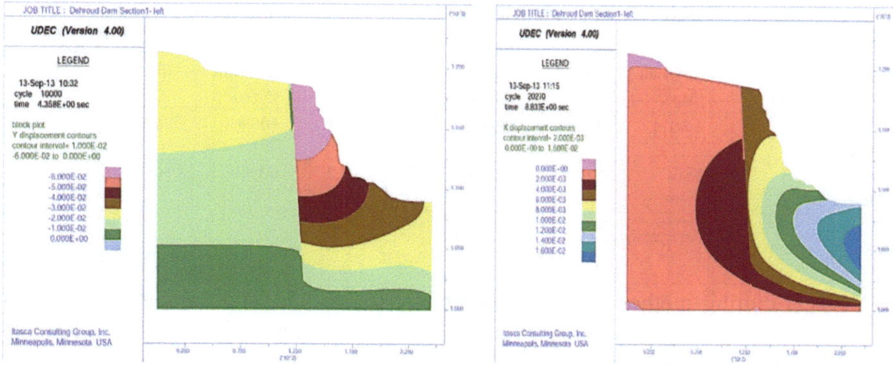

A: Mass displacement on x axis   B: Mass displacement on y axis

Fig. 7. The displacement after the fault F1 is considerably greater.

3.2 Stability Analysis the Left Cross-Section

As mentioned earlier, the rock mass in the left cross-section was ranked as poor to moderate according RMR value (35 to 50). Therefore, given the role of Fault F1 in this section and the structural analysis that shows there is no potential of structural loss, the stability analysis was performed based on the role of Fault F1 and the zone decomposed by it using UDEC Software.

The left and right sides of the fault have different mechanical and strength properties (Table 3). The properties determined for the left side of the fault by RMR are similar to the properties found in the right side. However, the properties related to the part after the fault toward the valley are different as evident in the reports in Table 4 (Phase II Report). These analyses were performed for the worst conditions i.e. the saturated and unsaturated conditions.

Table 3. Different mechanical and strength properties of the right and left sides of the abutment fault

Density (kg/m^3)	Transformability modulus (Pa)	Adhesion (Pa)	Friction angle	Poisson's ratio	Left abutment position
2250	3e9	60e3	17	12:28	Fault crushed zone
2350	4.22e9	150e3	20	12:26	Rock mass after to fault toward the valley
2530	20e9	1.6e6	32	12:23	Rock mass before the fault

3.3 Analysis of the Stability and Reliability of the Left Abutment of Haiqer Dam

As mentioned earlier, the rocks in the left section are classified as poor to moderate according to RMR value (35 to 50). Therefore, the continuous modeling was used

Table 4. Geotechnical parameters of rock components

In situ position	σ cm (dry/satu.) (Mpa)	c (dry/satu.) (Mpa)	φ (dry/satu.)	Ed (dry/satu (Gpa)	ν	G (dry/satu.) (Gpa)	σ ci (dry/satu.) (Mpa)	RMR (dry/satu.)	GSI (dry/satu.)
Right abutment	10.7	1.8/1.4	32/26	27/13	0.2	2.7/1.3	60/48	76/64	68/46
Left abutment	4.2/1	1/0.4	22/15	6.2/2.6	0.2	0.6/0.2	60/48	50/35	(47-37)/ 29

Vertical cutting methods are used optimally for circular rupture surfaces today in the equilibrium point analysis as circular shapes are suitable for analysis since they often have a close affinity to observed rupture shapes. However, it has been demonstrated that the precision of circular surfaces with slanted cutting, e.g. as in SARMA Method will result in more accurate responses in the analysis. Phase II Software and Reliability Analysis Modeling by Mathematica were used to perform a two-dimensional analysis of the wall (the left and right sides of fault have different mechanical and strength properties whose values were shown earlier in Table 3). Figure 8 shows the results of the two-dimensional analysis of the wall by Phase II Software.

Reliability Analysis Modeling of the left abutment of Haiqer Dam by Mathematica using SARMA Method shows similar results.

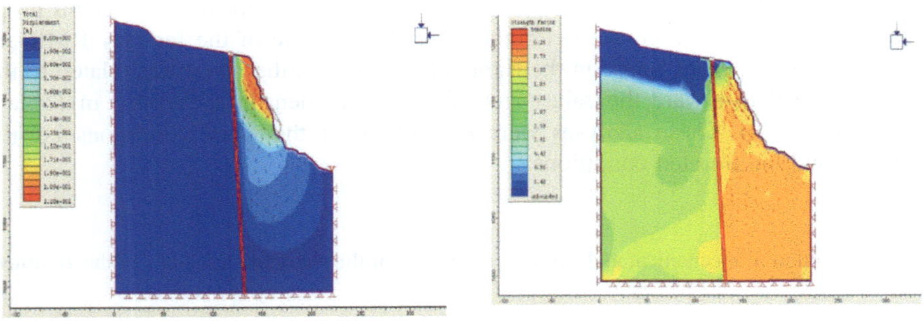

A: Conspicuous displacement in the right side

B: Lower safety factor in the ride side

Fig. 8. Results of the analysis of the left abutment of Haiqer Dam by Phase II Software

4 Conclusions

The analysis of stability and reliability of the slope of the walls, especially the slope of the jointed stones is one of the most significant issues taken into account when designing a dam. Such analysis can be done by geotechnical modeling. Accordingly, probabilistic methods can be employed to model the actual environment and function of natural systems such as rock and soil slopes through engineering experience, view, and judgment. Therefore, such methods including reliability analysis techniques provide better opportunities to apply professional theories under different circumstances to achieve acceptable solutions. Therefore, the following conclusions are made based on modeling and analyses performed in the course of the present study:

1. Models constructed in the Mathematica Software Environment are able to perform reliability analysis of jointed and highly decomposed rock slopes as evident in numerical methods.
2. Results related to the left abutment indicate that local loss is inevitable and the inundation of the water behind the dam will increase the depth of critical circles of such losses and a major part of the wall will be exposed to loss.
3. The right abutment is more stable and there is less possibility of considerable loss as there is not loss risk.
4. In general, the left abutment is less reliable than the right abutment so the former is exposed to a greater possibility of loss. Accordingly, the performance of more accurate assessments using observation methods can lead to more accurate and efficient results for stability and reliability analysis.
5. The results of reliability analysis using FORM Method and the model constructed through KGM for the left cross-section as well as the results from numerical modeling after the attainment of point equilibrium are indicative of a sliding surface. However, this surface does not exist in 3D form in reality so the traces of layering and joint set No. 3 perpendicular to this section were analyzed; suggesting that in general there is no probability of general planar loss for this section of the wall.
6. The 2D analysis of the left abutment using Phase II Software and reliability analysis modeling in Mathematica software environment through SARMA Method shows similar results.

Acknowledgements. This article was compiled based on a Master's Thesis at Islamic Azad University, Estahban Branch. The authors would like to thank the chancellor and staff of the university and also express their sincere thanks to Fars Regional Water Organization for its cooperation in providing geotechnical and geological information of Haiqer Dam.

References

Bray, J.W,. Goodman, R.E.: The theory of base friction models. In: International Journal of Rock Mechanics and Mining Sciences & Geomechanics Abstracts, vol. 18, No. 6, pp. 453–468. Pergamon (1981, December)

Goodman, R.E.: Block theory and its application. In: Daemen, J.J.K., Schultz, R.A. (eds.) Proceedings of 35th US Symposium on Rock Mechanics. AA Balkema. Rotterdam. 1995b. str, pp. 3–15 (1995)

Lin, D., Fairhurst, C., Starfield, A.M.: Geometrical identification of three-dimensional rock block systems using topological techniques. In: International Journal of Rock Mechanics and Mining Sciences & Geomechanics Abstracts, vol. 24, no. 6, pp. 331–338. Pergamon (1987, December)

Tonon, F., Bernardini, A., Mammino, A.: Reliability analysis of rock mass response by means of random set theory. Reliab. Eng. Syst. Saf. **70**(3), 263–282 (2000)

Warburton, P.M.: Vector stability analysis of an arbitrary polyhedral rock block with any number of free faces. In: International Journal of Rock Mechanics and Mining Sciences & Geomechanics Abstracts, vol. 18, no. 5, pp. 415–427. Pergamon (1981, October)

Wibowo, J.L.: Consideration of secondary blocks in key-block analysis. Int. J. Rock Mech. Min. Sci. **34**(3–4), 333-e1 (1997)

Yarahmadi Bafghi, A.R., Verdel, T.: The key-group method. Int. J. Numer. Anal. Meth. Geomech. **27**(6), 495–511 (2003)

Yarahmadi Bafghi, A.R., Verdel, T.: The probabilistic key-group method. Int. J. Numer. Anal. Meth. Geomech. **28**(9), 899–917 (2004)

An Enhanced Solution for the Expansion of Cylindrical Cavities in Modified Cam Clay

Vincenzo Silvestri[1(✉)] and Claudette Tabib[2]

[1] École Polytechnique, Montreal, Canada
vincenzo.silvestri@polymtl.ca
[2] Consultant, Montreal, Canada
claudette.tabib@yahoo.ca

Abstract. A solution was recently obtained for the expansion of cylindrical cavities in normally consolidated Modified Cam Clay under plane strain and undrained conditions. Constitutive relationships which combined the elastic and plastic responses of the clay were derived by assuming that the shear modulus, G', and the hardening parameter, p'_c, which controls the size of the yield locus, remained constant throughout the expansion. The latter assumption resulted in some unexpected findings. It was found, for example, that the ratio between the axial stress σ_{zf} and the sum of the remaining radial and tangential stresses, σ_{rf} and $\sigma_{\theta f}$ at critical state, closely approached but never attained the theoretical value of 0.5, except for the particular case of $K_o = 1$, where K_o is the coefficient of lateral earth pressure. The present paper removes the restrictive assumption on the hardening parameter, p'_c, which now varies during the shearing process. New effective stress relationships are obtained and the solution is applied for illustration purposes to both normally consolidated and overconsolidated Boston Blue Clay, for which the overconsolidation ratio (OCR) varies between 1 and 32. Results are compared to a well - known benchmark solution obtained by means of finite element analyses.

Keywords: Undrained cylindrical expansion · Plane strain · Modified cam clay
Stress distributions · Overconsolidated clay · Analytical and numerical analyses
Boston blue clay · Comparisons

1 Introduction

Results of analyses of expansion of vertical cylindrical cavities in clay are generally used for interpretation of pressuremeter tests and effects caused by installation of piles (Yu 2000). For many years, studies were carried out for soil idealized as either elastic perfectly plastic, or nonlinear plastic material. Such studies, which dealt with total stress approaches, were initiated by Gibson and Anderson (1961), and later expanded by Baguelin et al. (1972), Ladanyi (1972), and Palmer (1972). Investigations were also performed for soil represented as a work-hardening elastic plastic material, with properties based upon the effective stress model of Modified Cam Clay (see, for example, Carter et al. 1979; Randolph et al. 1979; Collins and Yu 1996; Yu 2000; and Cao et al. 2001).

H. Shehata and C. S. Desai (Eds.): GeoMEast 2018, SUCI, pp. 101–117, 2019.
https://doi.org/10.1007/978-3-030-01926-6_7

While the elastic perfectly plastic model has the advantage that a closed-form solution is possible for cylindrical cavity expansion, the model has two important shortcomings (Randolph et al. 1979). First, no account is taken of pore water pressure generated in pure shear. Second, the model lacks the capability of linking soil strength and its change with the current effective stress state and stress history. The Modified Cam Clay model has been used in this paper to overcome such shortcomings. The solution, obtained by Silvestri and Abou-Samra (2012) for the effective stresses generated during the undrained plane-strain expansion of a cylindrical cavity in normally consolidated Cam Clay, has been modified and applied in conjunction with a simple elastic model to the analysis of expansion tests in overconsolidated remoulded Boston Blue Clay.

Limit radial expansion and excess pore water pressures are determined for different values of overconsolidation ratio (OCR). Results are compared both with data reported by Randolph et al. (1979) which were obtained using a finite element analysis and with results obtained from finite difference simulations using the FLAC software (ltasca 1995).

2 Modified Cam Clay

A brief introduction to the Modified Cam Clay model is presented in this paper. Interested readers will find an in-depth treatment in Wood (2007). Application of Modified Cam Clay theory to cavity expansion problems is also extremely well covered in Yu (2000). The generalized stress and strain parameters used in Modified Cam Clay are the following:

$$p' = \frac{\sigma_1' + \sigma_2' + \sigma_3'}{3} \tag{1}$$

$$q = \left[\frac{(\sigma_1' - \sigma_2')^2 + (\sigma_2' - \sigma_3')^2 + (\sigma_3' - \sigma_1')^2}{2}\right]^{1/2} \tag{2}$$

$$\varepsilon_v = \varepsilon_1 + \varepsilon_2 + \varepsilon_3 \tag{3}$$

$$\varepsilon_q = \sqrt{2}\left[\frac{(\varepsilon_1 - \varepsilon_2)^2 + (\varepsilon_2 - \varepsilon_3)^2 + (\varepsilon_3 - \varepsilon_1)^2}{9}\right]^{1/2} \tag{4}$$

where p' = mean effective stress; q = deviator stress; ε_v = volumetric strain; ε_q = deviatoric strain; σ'_1, σ'_2, σ'_3 = principal effective stresses; and ε_1, ε_2, ε_3 = true principal strains.

The elliptical yield locus is given by:

$$q = p'M\left[(p_c'/p') - 1\right]^{1/2} \tag{5}$$

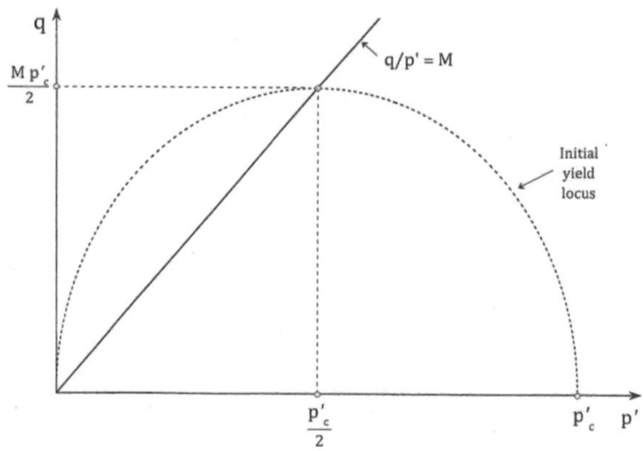

Fig. 1. Yield locus of Modified Cam Clay

where $M = q/p''$ at critical stress state and p'_c is a hardening parameter which controls the size of the yield locus, as shown in Fig. 1. The undrained effective stress path (ESP) followed by normally consolidated Modified Cam Clay is described by the following relationship (Wood 2007):

$$\eta = \frac{q}{p'} = \left[\left(M^2 + \eta_i^2\right) \left(p'_i / p'\right)^{1/\Lambda} - M^2 \right]^{1/2} \tag{6}$$

where p'_i and $\eta_i = q_i / p'_i$ define an initial anisotropic effective stress state (i.e., point A in Fig. 2); $\Lambda = (\lambda - \kappa)/\lambda$; λ = slope of normal consolidation line in $v : \ln p'$ space; κ = slope of swelling line in $v : \ln p'$ space; $v = 1 + e$ = specific volume with e = void ratio. The stress path begins at point A (p'_i, q_i) and ends at point B (p'_f, q_f) on the critical state line, whose coordinates are $p'_f = p'_i\, ((M^2 + \eta_i^2)/2M^2)^\Lambda$ and $q_f = M\, p'_f$.

Equation 6 reduces to

$$q = p' M \left[\left(p'_o / p'\right)^{1/\Lambda} - 1 \right]^{1/2} \tag{7}$$

when the initial stress state is isotropic, for $p'_i = p'_o$ and $q_i = 0$. As a consequence, the coordinates at point B are also given by $p'_f = 2^{-\Lambda}\, p'_o$ and $q_f = M\, p'_f = 2^{-\Lambda} M\, p'_o$. The maximum value of the deviator stress q equals $M\, p'_o\, (2\Lambda - 1)^{1-1/2\Lambda} / (2\Lambda)^\Lambda$ and corresponds to $p' = p'_o\, ((2\Lambda - 1)/2\Lambda)^\Lambda$, as shown in Fig. 2.

Equations 6 and 7 describe the undrained effective stress path (ESP) provided yielding and plastic volumetric strains are occurring, as for normally consolidated Modified Cam Clay (Wood 2007). For overconsolidated clay, when the effective stress state lies inside the current yield locus, the soil behaves in an elastic manner and the only way that a constant volume condition can be maintained during an undrained test is for the mean effective stress p' to remain constant (Schofield and Wroth 1968; Wood 2007). Consequently, for an initially overconsolidated clay that undergoes undrained

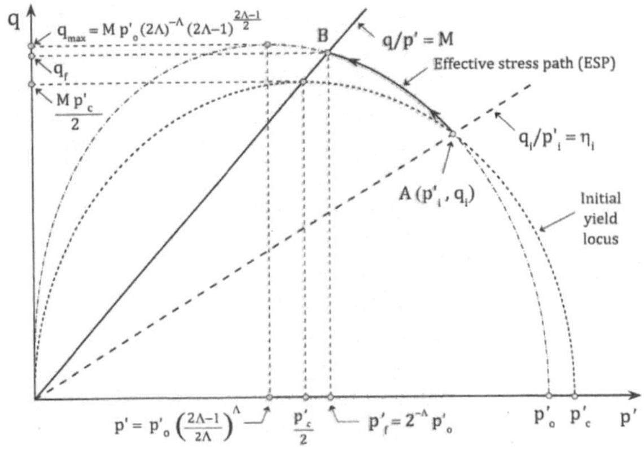

Fig. 2. Effective stress path of normally consolidated Modified Cam Clay

shearing, the effective stress path must rise at constant p' until the current yield locus is reached (i.e., the linear segment AB in Fig. 3) and the elastic plastic effective stress path of Eq. 6 can be joined (i.e., the curved segment BC in Fig. 3).

Changes in deviator stress q within the yield locus, for isotropic elastic soil, cause no change in volume but do produce elastic deviatoric strains that can be calculated from (Wood 2007):

$$\varepsilon_q = (q - q_i)/3G' \tag{8}$$

for $q < q_B$ in Fig. 3, where G' is the shear modulus.

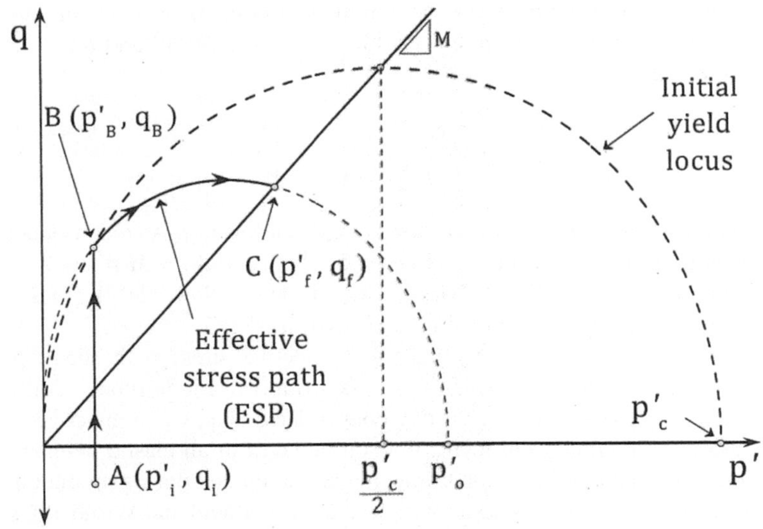

Fig. 3. Effective stress path of overconsolidated Modified Cam Clay

It is also assumed herein that the critical failure state is represented by a unique von Mises criterion with constant deviator stress q_f, provided that the specific volume is the same for normally overconsolidated and overconsolidated soil. As a result, it will theoretically be possible for heavily overconsolidated clay sheared undrained to sustain a deviator stress q in excess of q_f, as shown in Fig. 3 (see, also, Atkinson 1993). Such situation arises when the initial mean effective stress p'_i is less than p'_f.

3 Undrained Expansion of Vertical Cylindrical Cavity

3.1 Theoretical Analysis

It is convenient to consider cylindrical coordinates r, θ, z for the determination of the principal stresses around an expanding vertical cylindrical cavity. Let the total principal stresses in the radial, tangential and axial (vertical) directions in the clay surrounding the cavity be denoted by σ_r, σ_θ, σ_z, the corresponding effective stresses by σ'_r, σ'_θ, σ'_z, and the true principal strains in the same directions by ε_r, ε_θ, ε_z. The initial effective stresses existing in the soil are $\sigma'_{z\,i}$ and $\sigma'_{r\,i} = \sigma'_{\theta\,i} = K_o{}'_{z\,i}$, where K_o is the coefficient of lateral earth pressure at rest. The initial mean effective stress p'_i and deviator stress q_i are thus equal to $\sigma'_{z\,i}\,(1 + 2\,K_o)\,/3$ and $\sigma'_{z\,i}\,(1 - K_o)$, respectively. The ambient pore water pressure u_i is assumed to be zero.

As true strains are considered in the analysis, it is in order to distinguish the radial distance r of a material element in the unstrained condition in the medium and the corresponding distance r' of the same material element in the strained condition, similarly the cavity radius a before, and a' after the distortion has occurred.

The true tangential strain ε_θ is defined as (Nadai 1950; Yu 2000)

$$\varepsilon_\theta = ln\left(\frac{r'}{r}\right) = \frac{1}{2}ln[\left(\frac{r'}{r}\right)^2] \tag{9}$$

Because the expansion of the cavity takes place in undrained conditions and in plane strain in the vertical direction, it follows that $\varepsilon_r + \varepsilon_\theta = 0$, leading to $\varepsilon_r = -\varepsilon_\theta$, since $\varepsilon_r + \varepsilon_\theta + \varepsilon_z = 0$ and $\varepsilon_z = 0$. As a result, the shear strain γ mobilized in the horizontal plane, which is defined as $\gamma = |\varepsilon_r - \varepsilon_\theta|$ becomes equal to 2 ε_θ or,

$$\gamma = \ln[\left(\frac{r'}{r}\right)^2] \tag{10a}$$

The shear strain at the wall of the cavity is given by

$$\gamma = \gamma_a = \ln[\left(\frac{a'}{a}\right)^2] \tag{10b}$$

for $r' = a'$ and $r = a$. In addition, the shear stress τ in the horizontal plane equals either $(\sigma'_r - \sigma'_\theta)\,/2$ or $(\sigma_r - \sigma_\theta)\,/2$.

The total radial stress σ_r generated in the soil medium is given by (Nadai 1950; Yu 2000)

$$\sigma_r = \int_0^{\gamma} \frac{\tau d\gamma}{e^{\gamma} - 1} + \sigma_{ri} \tag{11}$$

Consequently, the total radial stress at the wall of the cavity becomes

$$\sigma_{ra} = \int_0^{\gamma_a} \frac{\tau d\gamma}{e^{\gamma} - 1} + \sigma_{ri} \tag{12}$$

The determination of the limit expansion pressure is made easier by the use of the Almansi tangential strain α (Baguelin et al. 1978)

$$\alpha = \frac{1}{2}\left(\frac{r'^2 - r^2}{r'^2}\right) \tag{13}$$

which reduces to

$$\alpha_o = \frac{1}{2}\left(\frac{a'^2 - a^2}{a'^2}\right) \tag{14}$$

at the wall of the cavity. Comparing Eqs. 13 and 10a shows that $\gamma = -\ln(1\text{-}2\alpha)$ or, inversely, that $\alpha = (e^{\gamma} - 1)/2e^{\gamma}$. As a result, Eqs. 11 and 12 may also be written as:

$$\sigma_r = \int_0^{\alpha} \frac{\tau d\alpha}{\alpha} + \sigma_{ri} \tag{15}$$

and

$$\sigma_{ra} = \int_0^{\alpha_o} \frac{\tau d\alpha}{\alpha} + \sigma_{ri} \tag{16}$$

The limit pressure at the wall of the cavity is obtained by letting $\alpha_o = 0.5$ in Eq. 16, because the shear strain γ_a tends to ∞ for $\alpha_o = 0.5$ or $a'/a = \infty$ in Eq. 14

Consideration of Eq. 16 indicates that the slope of the relationship between the total radial pressure and the logarithm of the Almansi tangential strain at the wall of the cavity directly gives the shear stress τ as a function of strain. Such point which was exploited by Gibson and Anderson (1961), Baguelin et al. (1972), Ladanyi (1972), and Palmer (1972), for the determination of the undrained shear strength of clay from pressuremeter test results, was already known in the field of the plastic expansion of thick - walled tubes (see, for example, Shepherd 1948; Hill 1950; Nadai 1950).

3.2 Elastic Phase of Expansion

It was mentioned above that the segment AB of the undrained effective stress path had to rise vertically at constant mean effective stress p' inside the current yield locus, as shown in Fig. 3. Consequently, no excess pore water pressure is generated during shearing along AB and the effective stresses σ'_r, σ'_θ, σ'_z, are thus given by (see, also, Baguelin et al. 1978):

$$\sigma'_r = \sigma'_{ri} + \Delta\sigma_r \tag{17}$$

$$\sigma'_\theta = \sigma'_\theta + \Delta\sigma_\theta \tag{18}$$

$$\sigma'_z = \sigma'_{zi} \tag{19}$$

where $\sigma'_{ri} = \sigma'_{\theta i} = K_o\sigma'_{zi}$, and $\Delta\sigma_r$ is an arbitrary radial stress increment, with $\Delta\sigma_r = -\Delta\sigma_\theta$. Replacement of Eqs. 17, 18, and 19 into Eq. 2 allows the determination of the increment $\Delta\sigma_r$ in terms of any deviator stress q between q_i and q_B. The increment is simply

$$\Delta\sigma_r = (q^2 - q_i^2)^{1/2}/\sqrt{3} \tag{20}$$

As a result, Eqs. 17 and 18 become

$$\sigma'_r = \sigma'_{ri} + (q^2 - q_i^2)^{1/2}/\sqrt{3} \tag{21}$$

$$\sigma'_\theta = \sigma'_{\theta i} + (q^2 - q_i^2)^{1/2}/\sqrt{3} \tag{22}$$

Thus the shear stress $\tau = (\sigma'_r - \sigma'_\theta)/2 = \Delta\sigma_r = (q^2 - q_i^2)/\sqrt{3}$.

The value of the deviator stress q_B at point B in Fig. 3 is found from Eq. 7 for $p'_B = p'_i$. Once q_B is obtained, the parameter p'_c which controls the size of the initial yield locus is computed from Eq. 5 for $p'_B = p'_i$ and $q = q_B$.

The soil behaves like a simple elastic material with constant shear modulus G' for undrained shearing along AB. Consequently, the shear stress τ is also given by

$$\tau = G'\gamma \tag{23}$$

3.3 Plastic Phase of Expansion

The soil becomes plastic and the path follows the curved segment BC, when the effective stress path reaches point B in Fig. 3. The coordinates of point C on the critical state line are, as before, $p'_f = 2^{-\Lambda} p'_o$ and $q_f = M p'_f = 2^{-\Lambda} M p'_o$. The stress path BC is described by Eq. 6 or 7.

For constant shear modulus G', the effective stresses σ'_r, σ'_θ, σ'_z, generated along BC in Fig. 3 may be determined using the following relationships (Silvestri and Abou-Samra 2012):

$$d\sigma'_r = -2G'd\varepsilon_\theta + dp'\left[1 + \frac{6G'\kappa(\sigma'_r - p')}{\upsilon M^2 p'(2p' - p'_c)}\right] \tag{24a}$$

$$d\sigma'_\theta = 2G'd\varepsilon_\theta + dp'\left[1 + \frac{6G'\kappa(\sigma'_\theta - p')}{\upsilon M^2 p'(2p' - p'_c)}\right] \tag{24b}$$

$$d\sigma'_z = dp'\left[1 + \frac{6G'\kappa(\sigma'_z - p')}{\upsilon M^2 p'(2p' - p'_c)}\right] \tag{24c}$$

Integration of Eq. 24c yields

$$\sigma'_z = p' + (\sigma'_{zi} - p'_i)e^{\frac{6G'\kappa}{\upsilon M^2}F(p',p'_c)} \tag{25}$$

where

$$F(p',p'_c) = \int_{p'_i}^{p'} \frac{dp'}{p'(2p' - p'_c)} \tag{26}$$

The effective radial and tangential stresses, σ'_r and σ'_θ, are found by substituting the expression for σ'_z from Eq. 25 into Eqs. 1 and 2, leading to:

$$\sigma'_r = p' - \frac{(\sigma'_{zi} - p'_i)}{2}e^{\frac{6G'\kappa}{\upsilon M^2}F(p',p'_c)} + \tau \tag{27a}$$

$$\sigma'_\theta = p' - \frac{(\sigma'_{zi} - p'_i)}{2}e^{\frac{6G'\kappa}{\upsilon M^2}F(p',p'_c)} - \tau \tag{27b}$$

where the shear stress $\tau = (\sigma'_r - \sigma'_\theta)/2$ acting in the horizontal plane is equal to:

$$\tau = \frac{1}{2}\left[\frac{4}{3}q^2 - 3(\sigma'_{zi} - p'_i)^2 e^{\frac{12G/\kappa}{\upsilon M^2}F(p',p'_c)}\right]^{1/2} \tag{28}$$

Remark 1

The shear modulus G' was considered to remain constant in the present paper (see also, Zytynski et al. 1978; Randolph et al. 1979; Sheng et al. 2000).

Remark 2

Silvestri and Abou-Samra (2012) integrated Eq. 25 by assuming that the hardening parameter p'_c remained constant during expansion, resulting in closed - form, albeit approximate, expressions for σ'_r and σ'_θ. However, the hardening parameter p'_c is variable and is function of the current effective mean stress p' (Wood 2007). Indeed, because the specific volume v of the soil is constant in undrained shearing, the change in volume between p' and p'_i on the unloading line of slope κ is equal to that between p'_o and p'_c along the normal consolidation line of slope λ, resulting in:

$$v - kln\left(\frac{p'_c}{p'}\right) = v - \lambda ln\left(\frac{p'_c}{p'_o}\right) \tag{29a}$$

or

$$p'_c = p'\left(p'_o/p'\right)^{1/\Lambda} \tag{29b}$$

Equation 29b may also be simply obtained by equating Eqs. 5 to 7.

The assumption made by Silvestri and Abou-Samra (2012) to the effect that p'_c remained constant is found to be approximately valid only when p'_c is close to p'_f. This occurs when the clay is lightly overconsolidated.

Remark 3

Chen and Abousleiman (2012) also obtained an exact semi-analytical solution for cylindrical cavity expansion in Modified Cam Clay. However, their effective stress relationships were not given in the form of the explicit expressions reported in Eqs. 25 and 27a, 27b .

Remark 4

When the initial stress state is isotropic (i.e., for $\sigma'_{r\,i} = \sigma'_{z\,i} = \sigma'_{\theta\,i} = p'_i$), Eqs. 25 and 27a, 27b show that

$$\sigma'_r = p' + q/\sqrt{3} \tag{30a}$$

$$\sigma'_z = p' \tag{30b}$$

$$\sigma'_r = p' - q/\sqrt{3} \tag{30c}$$

and

$$\tau = (\sigma'_r - \sigma'_\theta)/2 = q/\sqrt{3} \tag{30d}$$

at all times. In addition, $\sigma'_r = (\sigma'_r + \sigma'_\theta)/2$ in such case. This result was also pointed out by Chen and Abousleiman (2012), and Silvestri and Abou-Samra (2012).

Remark 5

Because the value of the integral $\int_{p'_i}^{p'} \frac{dp'}{p'(2p'-p'c)}$ in Eq. 26 is negative, the exponential term in Eqs. 25 and 27a, 27b decreases continuously during shearing and tends to zero at critical state when p' approaches p'_f. Indeed, Eq. 29b shows that $p'_c = 2\,p'_f$ in such case, because $p'_f = 2^{-\Lambda} p'_o$ as already mentioned. As a result, the effective stresses σ'_{rf}, σ'_{zf}, $\sigma'_{\theta f}$ and the shear stress τ_f at critical state become:

$$\sigma'_{rf} = p'_f + q_f/\sqrt{3} \tag{31a}$$

$$\sigma'_{zf} = p'_f \tag{31b}$$

$$\sigma'_{\theta f} = p'_f - q_f/\sqrt{3} \tag{31c}$$

and

$$\tau_f = q_f/\sqrt{3} \tag{31d}$$

as pointed out by Randolph et al. (1979). Further σ'_{zf} is always equal to $(\sigma'_{rf}+\sigma'_{\theta f})/2$ at critical state.

3.4 Shear Stress - Shear Strain Relationship

The shear stress - shear strain relationship for undrained shearing along BC in Fig. 3 is given by (Silvestri and Abou-Samra 2012)

$$\gamma = \frac{\tau}{G'} - \frac{6\kappa}{\upsilon M^2} \int_{p'_i}^{p'} \frac{\tau dp'}{p'(2p' - p'_c)} \tag{32}$$

Substitution of Eqs. 28 and 32 into Eqs. 11 and 12 allows the determination of the total radial stresses σ_r and $\sigma_{r\,a}$ generated along BC in Fig. 3. The radial stresses were computed numerically, because of the mathematical complexity of the resulting relations.

4 Application

The theory presented above was applied to the determination of the stresses generated in the medium surrounding a cylindrical cavity in overconsolidated remoulded Boston Blue Clay modelled as Modified Cam Clay. The soil parameters are those reported by Randolph et al. (1979) : $v = 1 + e = 2.16$, $\lambda = 0.15$, $\kappa = 0.03$, $\Lambda = (\lambda - \kappa)/\lambda = 0.8$, and $M = 1.2$. Additional parameters are reported in Table 1.

The soil was considered to be initially one-dimensionally consolidated with $K_o = 0.55$ and allowed to swell back after unloading from the maximum vertical pressure. Different initial conditions were assumed prior to the expansion process. Each case corresponds to an overconsolidation ratio defined, in the usual way, as the ratio of the past maximum vertical effective pressure $\sigma'_{z\,max}$ to the actual vertical effective stress $\sigma'_{z\,i}$ (i.e., OCR = $\sigma'_{z\,max}/\sigma'_{z\,i}$). The OCR varied from 1 to 32. For every case, the void ratio before cavity expansion was 1.16, resulting in an initial specific volume v =

Table 1. Boston Blue Clay parameters (after Randolph et al. 1979)

Case	OCR	K_o	G'/S_u	$G'/\sigma'_{z\,i}$	G'/p'_i
A	1	0.55	74	25	36
B	2	0.7	83	50	63
C	4	1	91	101	100
D	8	1.35	100	201	164
E	32	2.75	118	806	372

2.16. The choice of a unique value of v was made so that all soils would reach the same deviator stress at critical state q_f (Randolph et al. 1979).Thus, the undrained shear strength S_u would be the same for all cases and equal to that mobilized in plane strain. This implies that $S_u = q_f/\sqrt{3}$, as indicated by Eq. 31d.

Case A (OCR = 1) was used as reference. The initial in situ vertical effective stress $\sigma'_{z\,i}$ was assumed to be equal to 300 kPa for calculation purposes, resulting in $\sigma'_{r\,i} = \sigma'_{\theta\,i} = K_o\sigma'_{z\,i} = 165$ kPa, $p'_i = 210$ kPa, $q_i = 135$ kPa, and $p'_o = 257$ kPa. The initial value of $p'_c = 270$ kPa. The undrained stress path (ESP) shown in Fig. 4 starts at $p'_i = 210$ kPa, $q_i = 135$ kPa, and ends at $p'_f = 147.6$ kPa, $q_f = 177.1$ kPa, yielding $S_u = 102.3$ kPa. The shear modulus G' was computed from the data reported in the fourth column of Table 1. An average value of 7570 kPa was retained for G'.

The value of $S_u = 102.3$ kPa was considered to be applicable in all cases. Table 2 presents the remaining initial stress parameters which are based upon the ratios given in Table 1. The initial yield locus parameter p'_c for OCR > 1 was found by first computing the value of the deviator stress q_B at point B in Fig. 3. Consider, for example,

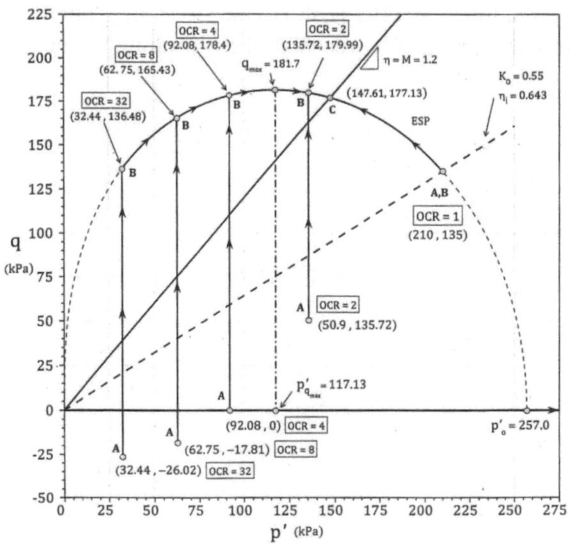

Fig. 4. Effective stress paths of all specimens

Table 2. Initial stress parameters

Case	OCR	$\sigma'_{z\,i}$ (kPa)	$\sigma'_{r\,i} = \sigma'_{\theta\,i}$ (kPa)	p'_i (kPa)	q_i (kPa)	G' (kPa)	p'_o (kPa)	p'_c (kPa)
A	1	300	165	210	135	7570	257	270
B	2	169.8	118.9	135.9	51	8488	257	301.4
C	4	92.2	92.2	92.2	0	9307	257	332.1
D	8	50.9	68.7	62.8	−17.8	10227	257	365.6
E	32	15.0	41.2	32.5	−26.2	12679	257	431.3

Case D (OCR = 8) for which p'_i = 62.8 kPa and q_i = - 17.8 kPa. Application of Eq. 7 gives q_B = 165.4 kPa for $p'_i = p'_B$ = 62.8 kPa and p'_o = 257 kPa. In addition, p'_c = 365.6 kPa from Eq. 5 for p'_i = 62.8 kPa and q_B = 165.4 kPa. The effective stress path is vertical between (p'_i = 62.8 kPa, q_i = − 17.8 kPa) and ($p'_i = p'_B$ = 62.8 kPa, q_B = 165.4 kPa). The remaining portion of the stress path is drawn in Fig. 4 using either Eq. 6 or 7. The parameter p'_o equals 257 kPa as for Case A, since the undrained effective stress path is unique. All the stress paths are grouped in Fig. 4. Examination of the stress paths for OCR > 1 shows that the maximum deviator stress q reached during shearing equals 181.7 kPa and is higher than q_f = 177.1 kPa. It occurs at p' = 117.3 kPa.

The effective stresses generated in the soil medium surrounding the cavity were determined in all cases. For comparison purposes, only the results obtained for Case D (OCR = 8) are reported in detail in the present paper. Figure 5 shows the distribution of the principal effective stresses and the excess pore water pressure generated in the soil at the instant when a'/a = 2. This ratio corresponds to a cavity shear strain γ_a = 1.386 or an Almansi tangential strain α_o = 0.375.

According to Randolph et al. (1979), the radial stress $\sigma_{r\,a}$ and the excess pore water pressure u would have practically reached their limiting values when a'/a = 2.

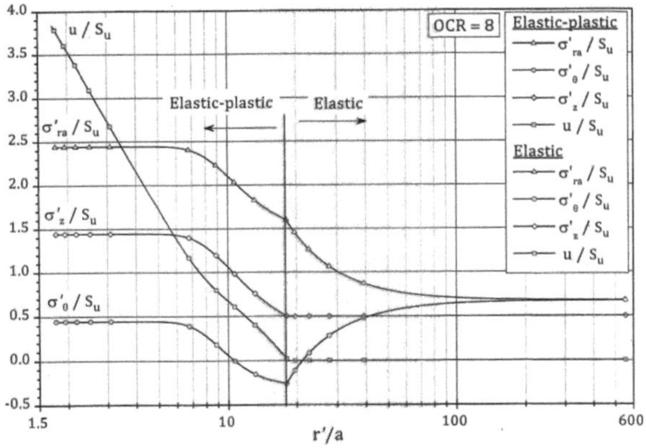

Fig. 5. Effective stress and excess pore pressure distributions around cylindrical cavity for a'/a = 2

Examination of the data given in Fig. 5 indicates that there exists a region of tensile effective tangential stress from $r'/a = 11$ to $r'/a = 18$. In addition, the stresses are elastic in the region that corresponds to r'/a in excess of 18.

Figure 6 presents, for comparison purposes, the results reported by Randolph et al. (1979) which were obtained using a finite element analysis. Corresponding results which were obtained using the finite difference code FLAC (Itasca 1995) were found to be essentially equivalent to those of Randolph et al. (1979) and are also given in the same figure. Examination of the data shows that there is good agreement with the results reported in Fig. 5. Again, there is a region of tensile effective tangential stress, located approximately between $r'/a = 11$ and $r'/a = 18$.

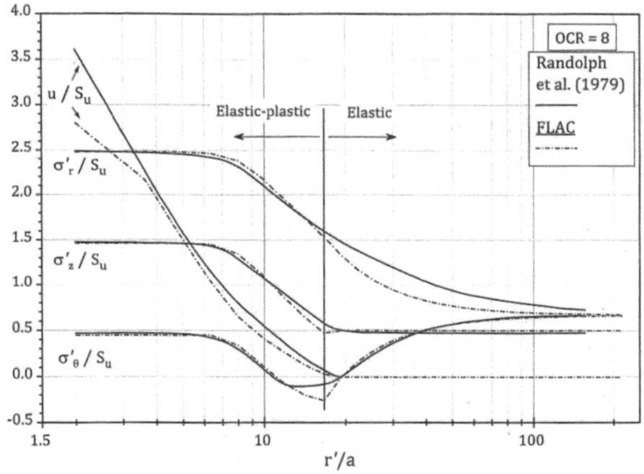

Fig. 6. Comparison between effective stress and excess pore pressure distributions

Figure 7 presents the normalized radial stress $\sigma_{r\ a}/S_u$ and excess pore water pressure u/S_u that are generated at the wall of the cavity, as function of the logarithm of the Almansi tangential strain α_o for case D (OCR = 8). The limiting values (i.e., for $\alpha_o = 0.5$) for $\sigma_{r\ a}$ and u are equal to 6.33 S_u and 3.88 S_u, respectively. Thus, the limiting excess radial pressure $\Delta\sigma_r$ equals 5.66 S_u, because $\sigma'_{r\ a} = 68.7$ kPa or 0.67 S_u. However, the excess radial and pore water pressures which correspond to $\alpha_o = 0.375$ or $a'/a = 2$ are equal to 5.37 Su and 3.6 S_u, respectively.

Table 3 summarizes the excess radial and pore water pressures computed for $a'/a = 2$ and compares them with the results reported by Randolph et al. (1979) and those determined using FLAC (ltasca 1995). Examination of the data indicates that the numerical results agree well with the theoretical values obtained in the present investigation.

Figure 8 shows the projections of the effective stress paths on the so-called π -plane, that is, the plane which is perpendicular to the space diagonal, $\sigma'_r = \sigma'_\theta = \sigma'_z$,

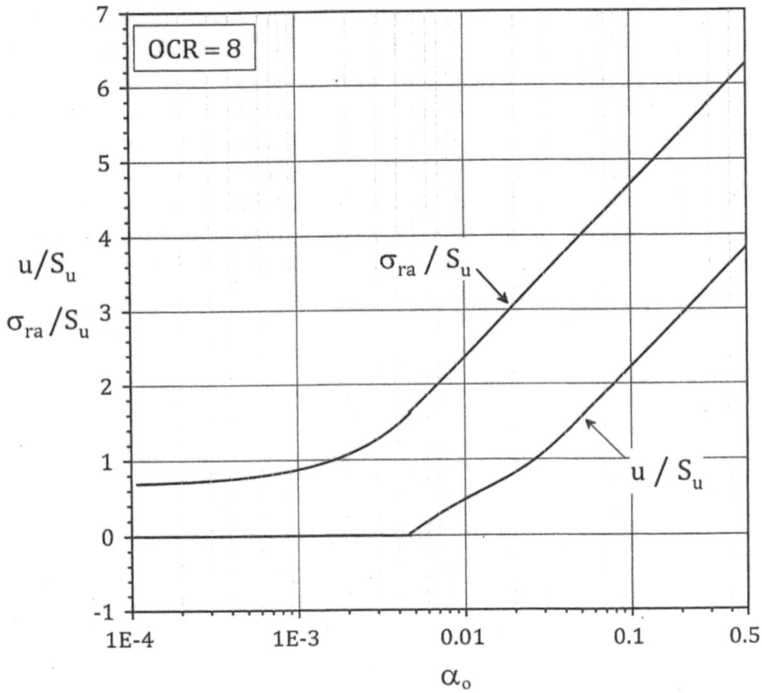

Fig. 7. Normalized radial stress and excess pore pressure generated at wall of cavity

in principal stress space (see, for example, Mendelson 1968). The coordinates x, y that are used to draw the stress paths are the following:

$$x = \frac{\sigma'_r - \sigma'_\theta}{\sqrt{2}} \tag{33a}$$

$$y = \frac{2\sigma'_z - \sigma'_r - \sigma'_\theta}{\sqrt{6}} \tag{33b}$$

There are several important observations that arise from the curves shown in this figure. First, all the stress paths end up on the circular von Mises criterion, as they should, since the value of the deviator stress at critical state q_f is the same in all cases. Second, stress paths for OCR $\geq$ 2 are partially located on the exterior of the von Mises criterion during the expansion process. Such response in caused by the fact that the deviator stress q can exceed q_f for overconsolidated clays, as shown in Fig. 3. Third, all the stress paths end up on the x-axis as shown by application of Eq. 33b since $y = 0$ for the stresses at critical state obtained from Eqs. 31a to 31c. In addition, the value of x at critical state, which equals $(\sigma'_{rf} - \sigma'_{\theta f})/\sqrt{2}$, reduces to $\sqrt{2}\,\tau_f = \sqrt{2}\,S_u = 144.6$ kPa.

Table 3. Comparison of excess radial stresses and pore pressures for $a''/a = 2$

		$\Delta\, \sigma_{r\ a} / S_u$			u / S_u		
Case	OCR	*1	*2	*3	*1	*2	*3
A	1	4.84	4.55	5.03	3.93	3.81	4.21
B	2	5.07	4.85	5.2	3.66	3.52	3.01
C	4	5.18	5.01	5.3	3.59	3.52	3.7
D	8	5.27	5.05	5.37	3.44	3.23	3.6
E	32	5.33	5.07	5.48	3.09	2.98	3.44

*1: Randolph et al. (1979)
*2: FLAC
*3: This study

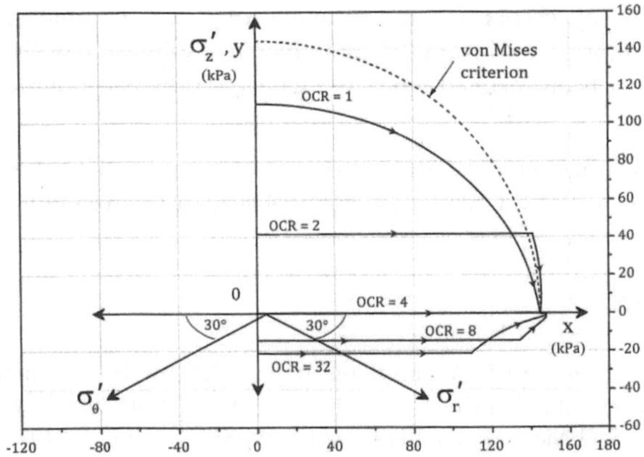

Fig. 8. Projections of effective stress paths on π-plane

Table 4 presents the total stresses at critical state $\sigma_{r\ f}$, $\sigma_{\theta f}$, $\sigma_{z\ f}$ and as well as the stress ratio $\sigma_{z\ f}/(\sigma_{r\ f}+\sigma_f)$ Examination of the data reported in the last column shows that the stress ratio equals 0.5, as it should.

Table 4. Total stresses at critical state

Case	OCR	$\sigma_{z\ f}$ (kPa)	$\sigma_{r\ f}$ (kPa)	σ_f (kPa)	$\sigma_{z\ f}/(\sigma_{r\ f}+\sigma_f)$
A	1	607	709	504	0.500
B	2	577	680	475	0.500
C	4	556	658	453	0.500
D	8	545	647	443	0.501
E	32	528	631	427	0.499

5 Conclusions

The following conclusions are drawn on the basis of the theoretical approach presented in this paper:

1. Improved principal effective stress relationships are obtained for the plane strain undrained expansion of vertical cylindrical cavities in Modified Cam Clay, by allowing the hardening parameter p'_c to vary during shearing.
2. Effective stress paths and stress distributions are obtained for specimens of normally consolidated and overconsolidated remolded Boston Blue Clay, modelled as Modified Cam Clay.
3. Comparisons with numerical results obtained by means of a finite difference code and with published results derived by means of a finite element solution show good agreement with the present variable - p'_c approach.
4. It is also shown by comparing the ultimate total stresses obtained in the present approach with those found from the finite element solution that the final total stress state had not yet been reached when the cavity radius had doubled in size.

Acknowledgements. The authors express their gratitude to the Natural Sciences and Engineering Research Council of Canada for the financial support received in this study.

REFERENCES

Atkinson, J.: The Mechanics of Soils and Foundations. McGraw-Hill, London (1993)

Baguelin, F., Jézéquel, J.F., Lemée, E., LeMéhauté, A.: Expansion of cylindrical probes in cohesive soils. J. Soil Mech. Found. Div. ASCE **98**(SM 11), 1129–1142 (1972)

Baguelin, F., Jézéquel, J.F., Shields, D.H.: The pressuremeter and foundation engineering. Trans Tech Publications, Clausthal (1978)

Cao, L.F., Teh, C.I., Chang, M.-F.: Undrained cavity expansion in modified Cam clay I: theoretical analysis. Géotechnique **51**(4), 323–334 (2001)

Carter, J.P., Randolph, M.F., Wroth, C.P.: Stress and pore pressure changes in clay during and after the expansion of a cylindrical cavity. Int. J. Anal. Numer. Methods Geomech. **3**(4), 305–322 (1979)

Chen, S.L., Abousleiman, Y.N.: Exact undrained elasto-plastic solution for cylindrical cavity expansion in modified Cam clay soil. Géotechnique **62**(5), 447–456 (2012)

Collins, L.F., Yu, H.S.: Undrained cavity expansions in critical soils. Int. J. Anal. Numer. Methods Geomech. **20**(7), 489–516 (1996)

Gibson, R.E., Anderson, W.F.: In-situ measurement of soil properties with the pressuremeter. Civil Eng. Public Works Rev. **56**(658), 615–618 (1961)

Hill, R.: The Mathematical Theory of Plasticity. Oxford University Press, London (1950)

Itasca Fast Lagrangian Analysis of Continua FLAC. Version 3.3. Itasca Consulting Group, Inc., Minneapolis (1995)

Ladanyi, B.: In-situ determination of undrained stress-strain behaviour of sensitive clays with the pressuremeter. Can. Geotech. J. **9**(3), 313–319 (1972)

Mendelson, A.: Plasticity: Theory and Application. MacMillan, New York (1968)

Nadai, A.: Theory of Flow and Fracture of Solids, vol. 1. McGraw-Hill, New York (1950)

Palmer, A.C.: Undrained plane-strain expansion of a cylindrical cavity in clay: a simple interpretation of the pressuremeter test. Géotechnique **22**(3), 451–457 (1972)
Randolph, M.F., Carter, J.P., Wroth, C.P.: Driven piles in clay-the effect of installation and subsequent consolidation. Géotechnique **24**(4), 361–393 (1979)
Schofield, A.N., Wroth, C.P.: Critical State Soil Mechanics. McGraw-Hill, London (1968)
Sheng, D., Sloan, S.W., Yu, H.S.: Aspects of finite element implementation of critical state models. Comput. Mech. **26**(2), 185–196 (2000)
Shepherd, A.N.: Plastic stress-strain relations. Proc. Inst. Mech. Eng. **159**, 95–99 (1948)
Silvestri, V., Abou-Samra, G.: Analytical solution for undrained plane strain expansion of a cylindrical cavity in modified Cam clay. Geomech. Eng. **4**(1), 19–37 (2012)
Wood, D.M.: Soil Behaviour and Critical State Soil Mechanics. Cambridge University Press, Cambridge (2007)
Yu, H.S.: Cavity Expansion Methods in Geomechanics. Kluwer Academic Publishers, Dordrecht (2000)
Zytynski, M., Randolph, M.F., Nova, R., Wroth, C.P.: On modelling the unloading-reloading behaviour of soils. Int J Anal. Numer. Methods Geomech. **2**(1), 87–93 (1978)

2D Spatial Variability Analysis of Sugar Creek Embankment: Comparative Study

Brigid Cami[1](✉), Sina Javankhoshdel[1], Thamer Yacoub[1], and Richard J. Bathurst[2]

[1] Rocscience Inc, Toronto, Canada
{brigid.cami,sina.javankhoshdel,thamer.yacoub}@rocscience.com

[2] GeoEngineering Centre at Queen's-RMC, Department of Civil Engineering, Royal Military College of Canada, Ontario, Canada
bathurst-r@rmc.ca

Abstract. Non-circular Random Limit Equilibrium Method (RLEM) uses a combination of non-circular limit equilibrium method, 2D random field theory, and Monte Carlo simulation. Non-circular RLEM has the ability to seek out the weakest failure path in a slope stability analysis using optimization techniques such as Surface Altering, while considering spatial variability of soil properties. This study is a probabilistic analysis of the stability of the Sugar Creek embankment fill slope using non-circular RLEM. This example has been investigated in the literature with other probabilistic methods such as the non-circular limit equilibrium method together with 1D spatial variability analysis. The results of this study show that the values of probability of failure are in good agreement for the 1D and 2D spatial variability cases, while the mean factor of safety values for 2D spatial variability tend to be higher with smaller standard deviations of factor of safety.

1 Introduction

This study focuses on the Sugar Creek embankment fill slope. The purpose of the study is to perform a probabilistic analysis with 2D spatial variability of the slope, and compare the results to the 1D spatial variability computations of Cho (2007).

Probabilistic stability analysis results considering spatial variability of soil properties and using limit equilibrium methods (LEM) have been reported in studies by Li and Lumb (1987), El-Ramly et al. (2001), Low (2003), Babu and Mukesh (2004), Cho (2007 and 2010), Hong and Roh (2008), Wang et al. (2011), Ji et al. (2012), Tabbaroki et al. (2013), Li et al. (2014), Javankhoshdel and Bathurst (2014) and Javankhoshdel et al. (2017).

There are few studies that investigate the influence of spatial variability of soil properties for the case of layered slopes. Huang et al. (2010) and Cho (2007) investigated the influence of spatial variability on probability of failure (P_f) using the Random Finite Element Method (RFEM) and the non-circular Random Limit Equilibrium Method (RLEM), respectively.

H. Shehata and C. S. Desai (Eds.): GeoMEast 2018, SUCI, pp. 118–125, 2019.
https://doi.org/10.1007/978-3-030-01926-6_8

Cho (2007) studied the Sugar Creek embankment fill slope, summarized the statistical properties of the slope, and computed limit equilibrium analyses using Spencer's method with 1D spatial variability along the slip surface, and Latin Hypercube sampling. This paper reports the results of analyses of the same slope with 2D spatial variability using the program *Slide* 2018 (Rocscience, 2018) and RLEM which is explained in the following section.

1.1 Random Limit Equilibrium Method (RLEM)

In the RLEM, a random field is first generated using the local average subdivision (LAS) method developed by Fenton and Vanmarcke (1990) and then mapped onto a grid of elements (mesh). Each mesh element in the random field has different values of soil properties, and cells close to one another have values that are closer in magnitude, based on the value of the spatial correlation length. In each realization, a search is carried out to find the mesh elements intersected by the slip surface. The random soil property values are assigned to the slices whose base mid-point falls within that element. A limit equilibrium approach (Morgenstern-Price Method) is then used to calculate factor of safety (F_S) for each realization. The probability of failure is calculated as the ratio of the number of simulations resulting in $F_S < 1$ to the total number of simulations.

The non-circular RLEM used in this study is a combination of Auto Refine search and the LEM (Morgenstern-Price method). The Auto Refine search is based on circular surfaces that are converted to piece-wise linear surfaces. The search for the lowest factor of safety is refined as the search progresses. An iterative approach is used, so that the results of one iteration, are used to narrow the search area on the slope in the next iteration.

The Auto Refine search in this study was used together with the Surface Altering optimization technique (Rocscience, 2018). When used in conjunction with a non-circular search this optimization method can be very effective and efficient at locating (searching out) slip surfaces with lower safety factors.

In this study, the Morgenstern-Price limit equilibrium method was used with the half sine interslice force function to calculate factor of safety. The bottom of each slice was sub-sampled at ten increments, and the values averaged, making it possible to use only 50 slices for the probabilistic analysis. A total of 50,000 Monte Carlo simulations were used for all probabilistic analyses.

1.2 Comparison to 1D Spatial Variability

1D spatial variability describes the variability occurring only along the slip surface of the slope. The same LAS method is used, but in one dimension, along the surface. The midpoint method is used to discretize the field, meaning that the field in each segment is described by a single random variable. As such, the field consists of the values at the centroid of each slice base.

Limit equilibrium methods and search methods can be applied in the same manner as traditional LEM. The Spencer Method and Latin-Hypercube sampling were used by Cho (2007).

2 2D Spatial Variability Analysis

2.1 Sugar Creek Embankment Fill Slope

The Sugar Creek embankment fill slope is reported in White et al. (2005). Cho (2007) created a typical slope section of the embankment and summarized the probabilistic properties of the slope. The same probabilistic parameters are used in this paper. Table 1 outlines the mean and coefficient of variation (COV = standard deviation/mean) values for friction angle (ϕ), cohesion (c), and unit weight (γ). Unit weight and pore pressure are considered to be deterministic in this analysis. Lognormal distributions are assumed for all variables, and cross-correlation between parameters is not considered, in order to best compare with the 1D spatial case. The slope section used in this study is shown in Fig. 1.

Table 1. Probabilistic soil properties of Sugar Creek embankment

Soil	ϕ (Peak friction angle) degrees		C (Cohesion) kPa		γ (Unit weight) kN/m^3
	Mean	COV	Mean	COV	Mean
Compacted fill	12.0	0.2	29	0.22	20.4
Alluvium	16.5	0.21	33.0	0.62	19.0
Highly weathered shale	12.8	0.38	33.2	0.60	20.0
Moderately weathered shale	21.6	0.44	97	1.38	20.0
Slightly weathered shale	23.3	0.48	675	1.86	21.0

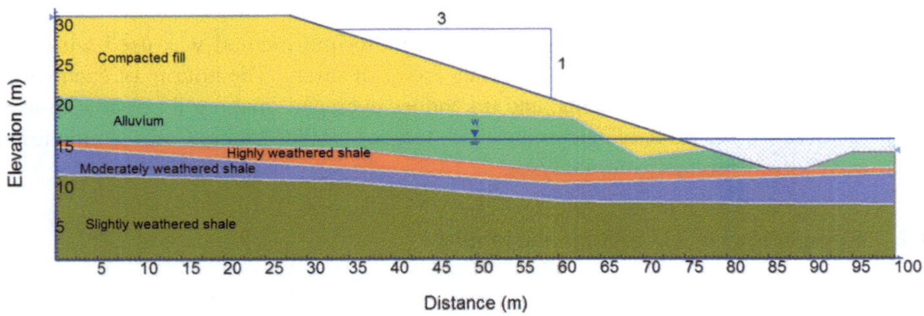

Fig. 1. Cross-section of the Sugar Creek embankment

As the correlation length is not known for this slope, this study considers three different isotropic correlation lengths of 20 m, 40 m, and infinity, in order to compare with Cho (2007).

2.2 Results

2.2.1 Deterministic Analysis: LEM and Shear Strength Reduction

The deterministic factor of safety was found to be 1.59 using the Morgenstern-Price method, the slip surface of which is shown in Fig. 2. This model was also computed using the shear strength reduction method as a check on the limit equilibrium method results. This resulted in a shear strength factor of 1.61 and is shown in Fig. 3. Both surfaces are in good agreement with the surface found in Cho (2007) using Spencer's method, as well as with the factor of safety of 1.62 found there.

2.2.2 RLEM

Using the RLEM, 50,000 simulations were computed for each correlation length case. Figure 4 shows the random field of cohesion in one of the 50,000 simulations for the 40 m isotropic correlation length case. This is the case that resulted in the lowest F_S value of 0.80 from the 50,000 samples. The total of these samples with F_S less than 1 are divided by 50,000 to obtain $\dot{P}_f$. It should be noted that the contours show relative standard deviation, meaning each color represents the increase or decrease in

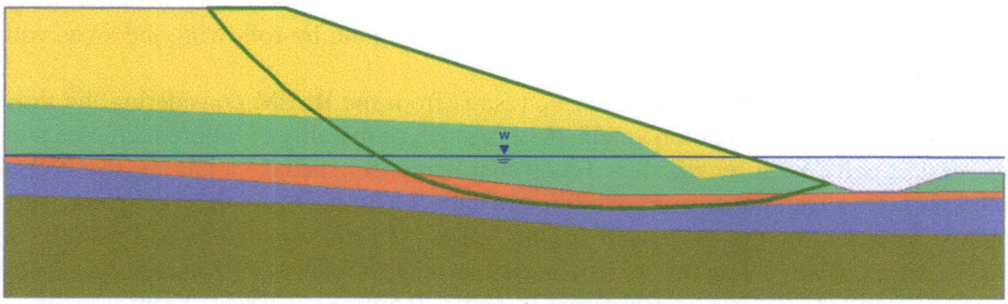

Fig. 2. Deterministic critical surface with $F_S = 1.59$

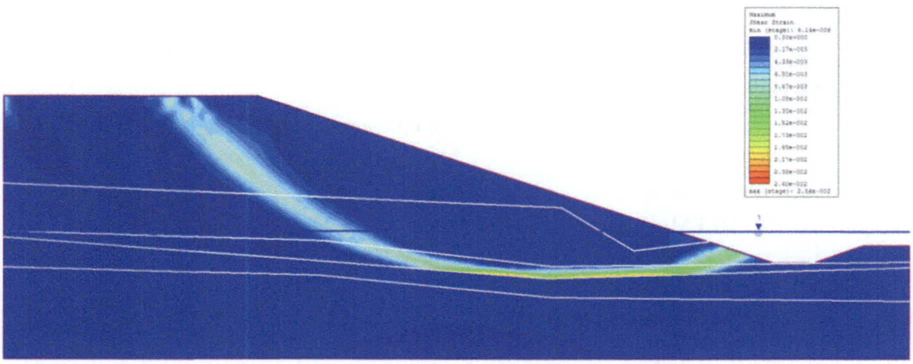

Fig. 3. Shear strength reduction: contours of maximum shear strain. Critical strength factor of 1.61. Computed with RS^2 (Rocscience, 2014)

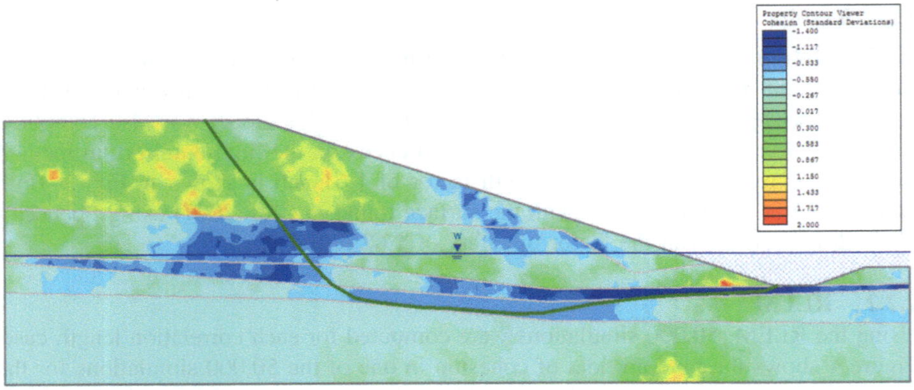

Fig. 4. Random field of cohesion for 40 m isotropic correlation length case. Minimum F_S simulation shown, with $F_S = 0.80$. Contours of relative standard deviation from mean cohesion displayed

increments of standard deviation from the mean cohesion value for each material. It can be seen that the surface shown seeks out the weakest path by following the areas with lower cohesive strength (bluer color).

Table 2 summarizes P_f and mean F_S results from the RLEM considering the three cases with different correlation length. As expected, probability of failure increases with increasing correlation length. It is found that using 2D spatial variability resulted in more conservative (greater) probabilities of failure compared to the 1D spatial variability analyses for the 20 m and infinity cases (0.13% vs. 0.12%, and 2.00% vs. 1.18%, respectively). For the 40 m correlation length case, the 2D spatial variability analysis resulted in a less conservative P_f value (0.59% vs. 0.88%). From a practical point of view the computed probabilities of failure using 2D and 1D spatial analyses are generally in good agreement.

Table 2. Computed 2D RLEM results and 1D spatial variability results (Cho, 2007), for each correlation length case

	Isotropic correlation length (m)		
	20 m	40 m	∞
2D RLEM P_f	0.13%	0.59%	2.00%
2D RLEM mean F_S	1.43 ($\sigma = 0.16$)	1.42 ($\sigma = 0.19$)	1.43 ($\sigma = 0.23$)
1D spatial variability P_f	0.12%	0.88%	1.18%
1D spatial variability mean F_S	1.63 ($\sigma = 0.28$)	1.63 ($\sigma = 0.35$)	1.63 ($\sigma = 0.39$)

The mean F_S values are essentially constant for each set of 2D and 1D spatial variability cases but each set of F_S values are different (e.g., about 1.4 and 1.6 for 2D and 1D spatial variability cases, respectively).

An interesting difference between 1D and 2D spatial variability analysis outcomes is the spread of F_S values. The 2D F_S values have a much smaller range, with almost half the standard deviation of 1D spatial variability cases (Table 2). This is interpreted to mean that 2D spatial variability analyses result in a more precise estimate of P_f.

Spatial variability of soil properties is generally an averaging process of soil properties over spatial correlation length. This averaging process reduces the level of uncertainty in the soil shear strength (decreasing the standard deviation and making the F_S distribution narrower) (Fenton and Griffiths, 2008). The final COV (standard deviation /mean) of F_s goes down since the influence of reducing the standard deviation is dominant. Thus, considering spatial variability reduces probability of failure.

Figure 5 shows a histogram of the factor of safety values for each correlation length case using the RLEM. As correlation length decreases, the spread of factor of safety values decrease as previously explained. The same trend was observed by Cho (2007) using the 1D spatial variability method.

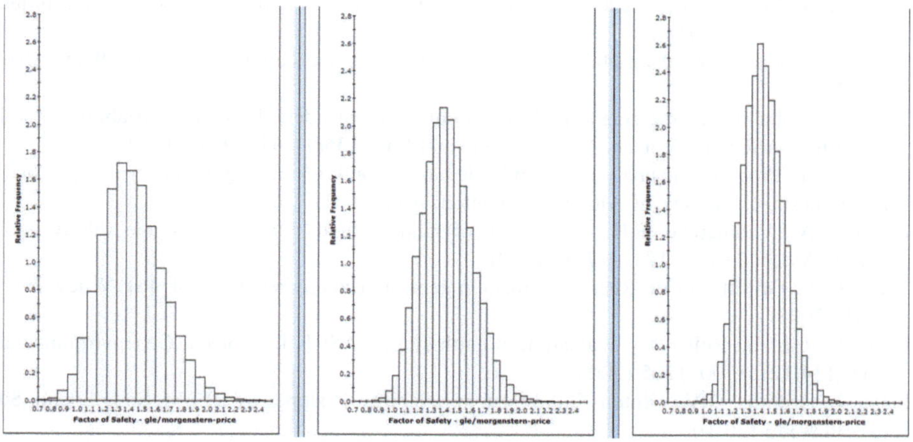

Fig. 5. Histogram of F_S for the correlation length cases of infinity, 40 m, and 20 m, respectively

3 Conclusions

This study focuses on the Sugar Creek embankment fill slope. The purpose of the study is to perform a probabilistic analysis with 2D spatial variability on the slope, and compare the results to the 1D spatial variability computation results of Cho (2007).

The P_f values were shown to increase with correlation length as expected. The results were practically the same using the 1D and 2D spatial variability methods. In all cases, P_f values were greater than the value of 0.01% recommended for design of embankments (Silva et al. 2008). The mean F_S values were essentially the same for each set of 1D and 2D spatial variability analyses. The mean factor of safety was approximately 1.4 using 2D spatial variability, compared to approximately 1.6 using 1D spatial variability.

Histograms of the factor of safety values for each correlation length case using the RLEM, showed that the spread in F_S values decreases with correlation length. The same trend was observed by Cho (2007).

As mentioned by Javankhoshdel et al. (2017), 1D spatial variability analysis is not as realistic as 2D spatial variability analysis. In 2D spatial variability analyses, different 2D random fields are generated and then probabilistic analyses are carried out. However, with 1D spatial variability analyses, spatial variability is confined to the slip surface which is divided into segments based on the value of the spatial correlation length. The spread in F_S values is also much smaller using 2D spatial variability analyses, indicating reduced uncertainty in the estimate of margin of safety using 2D spatial variability compared to 1D spatial variability analyses. Thus, the results of 2D spatial variability analysis are judged to be more reliable.

References

Babu, G.L.S., Mukesh, M.D.: Effect of soil variability on reliability of soil slopes. Geotechnique **54**(5), 335–337 (2004)

Cho, S.E.: Effects of spatial variability of soil properties on slope stability. Eng. Geol. **92**, 97–109 (2007)

Cho, S.E.: Probabilistic assessment of slope stability that considers the spatial variability of soil properties. ASCE J. Geotech. Geoenvironmental Eng. **136**(7), 975–984 (2010)

El-Ramly, H.: Probabilistic analyses of landslide hazards and risks: bridging theory and practice. Ph.D. thesis, University of Alberta, Edmonton, Alta (2001)

Fenton, G.A., Vanmarcke, E.H.: Simulation of random fields via local average subdivision. J. Eng. Mech. **116**(8), 1733–1749 (1990)

Fenton, G.A., Griffiths, D.V.: Risk assessment in geotechnical engineering, vol. 461. Wiley, New York (2008)

Hong, H., Roh, G.: Reliability evaluation of earth slopes. ASCE J. Geotech. Geoenvironmental Eng. **134**(12), 1700–1705 (2008)

Huang, J., Griffiths, D.V., Fenton, G.A.: System reliability of slopes by RFEM. Soils Found. **50** (3), 343–353 (2010)

Javankhoshdel, S., Bathurst, R.J.: Simplified probabilistic slope stability design charts for cohesive and c-ϕ soils. Can. Geotech. J. **51**(9), 1033–1045 (2014)

Javankhoshdel, S., Luo, N. Bathurst, R.J.: Probabilistic analysis of simple slopes with cohesive soil strength using RLEM and RFEM. Georisk (online) (2017)

Ji, J., Liao, H.J., Low, B.K.: Modeling 2D spatial variation in slope reliability analysis using interpolated autocorrelations. Comput. Geotech. **40**, 135–146 (2012)

Li, K.S., Lumb, P.: Probabilistic design of slopes. Can. Geotech. J. **24**, 520–535 (1987)

Li, D.Q., Qi, X.H., Phoon, K.K., Zhang, L.M., Zhou, C.B.: Effect of spatially variable shear strength parameters with linearly increasing mean trend on reliability of infinite slopes. Struct. Saf. **49**, 45–55 (2014)

Low, B.K. (2003). Practical probabilistic slope stability analysis. In: Proceedings, Soil and Rock America 2003, 12th Panamerican Conference on Soil Mechanics and Geotechnical Engineering and 39th U.S. Rock Mechanics Symposium, M.I.T., Cambridge, Massachusetts, vol. 2, pp. 2777–2784. Verlag Glückauf GmbH Essen, 22–26 June 2003

Rocscience. RS2 Version 9.0 – 2D finite element analysis. www.rocscience.com, Toronto, Ontario, Canada (2014)
Rocscience. Slide version 2018 – 2D limit equilibrium slope stability analysis. www.rocscience.com, Toronto, Ontario, Canada (2018)
Silva, F., Lambe, T.W., Marr, W.A.: Probability and risk of slope failure. ASCE J. Geotech. Geoenvironmental Eng. **134**(12), 1691–1699 (2008)
Tabarroki, M., Ahmad, F., Banaki, R., Jha, S.K., Ching, J.: Determining the factors of safety of spatially variable slopes modeled by random fields. ASCE J. Geotech. Geoenvironmental Eng. **139**(12), 2082–2095 (2013)
Wang, Y., Cao, Z.J., Au, S.K.: Practical reliability analysis of slope stability by advanced Monte Carlo simulations in a spreadsheet. Can. Geotech. J. **48**(1), 162–172 (2011)
White, D.J., Schaefer, V.R., Yang, H. Thompson, M.J.: Innovative solutions for slope stability reinforcement and characterization: vol. I, Final report CTRE Project 03-127. Center for Transportation Research and Education Iowa State University (2005)

The Influence of Rayleigh Coefficients on Numerical Calculation in Specific Dynamic Problems

Anna Borowiec(✉)

AGH University of Science and Technology, Cracow, Poland
anabo@agh.edu.pl

Abstract. The paper deals with the problem of choosing proper Rayleigh coefficients and their influence on numerical calculations in Soil Dynamics. Numerical analysis (Finite Element Method here) is conducted for two selected problems and calculated in *Plaxis*. In the first case the underground structure with the impulse wave applied to the subsoil is considered. The second model describes vibro-isolation incorporated into the generator foundation in purpose to protect the neighboring building. In both problems the wave propagation in porous medium and as a consequence the proper choice of Rayleigh coefficients takes major part. In purpose of drawing any decisive conclusions different physical values are investigated. The change in Rayleigh damping factors influences not only the specific value of result but also the behavior of the model as whole.

1 Introduction

General problem with establishing dynamic characteristic of soils derives from the fact that soil as a material is shapeless (in the contrary to steel or wood or concrete), what makes the determination of basic dynamic features (like modes of free vibrations) hard or even impossible. On the other hand the damping capability of the ground is introduced more naturally by the soil plasticity and (in most numerical programs) the fact that the wave is absorbed by the model's absorbent boundary. Still the need of use of Rayleigh damping coefficients (α and β) occurs for many problems, for example in case of diverse lithology composed of different soil layers of the small thicknesses.

The paper deals with the problem of choosing proper coefficients *alpha* and *beta* while solving two specific cases: underground structure (shelter) subjected to the overpressure wave and vibro-isolation of the machine foundation. Models were created and calculated by means of *Plaxis* computation system. In both problems dependency of damping characteristics on frequency (not on the shear strain) is considered thus the knowledge of the excitation or response frequencies is crucial.

1.1 Rayleigh Damping Coefficient Dependency on Frequencies

Firstly preliminary analysis of damping coefficients for different types of soil is performed using dynamic characteristics well known in situ and laboratory test. Figure 1

H. Shehata and C. S. Desai (Eds.): GeoMEast 2018, SUCI, pp. 126–132, 2019.
https://doi.org/10.1007/978-3-030-01926-6_9

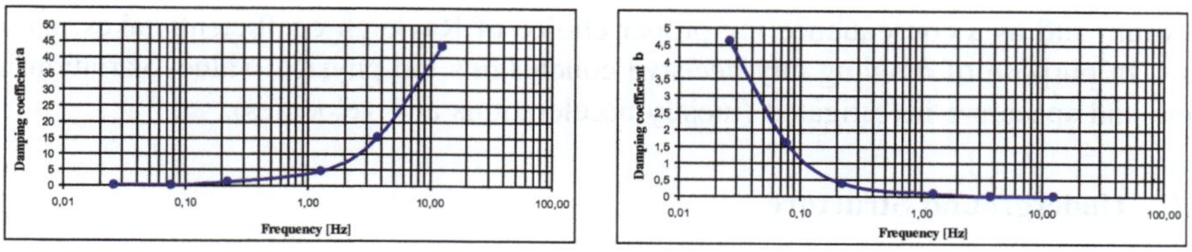

Fig. 1. Rayleigh damping coefficients vs. frequency for Sandy Clay.

shows correlation between damping coefficients α, β and frequency of excitation f [Hz] for Sandy Clay. In general Rayleigh coefficients are obtained from damping ratios ξ using a relationship that can be written as:

$$\alpha + \beta\omega_i^2 = 2\xi_i\omega_i \tag{1}$$

To calculate *alpha* and *beta* from Eq. (1) two damping ratios ξ_i for two given radial frequencies ω_i are required so that the system of linear equations can be formed. The procedure was carried out for several cohesive and non-cohesive soils and results for sands are presented at Fig. 2. Damping ratios ξ_i are adopted after Mitchell (2005) and Lin (1996) where they were are obtained from laboratory tests. They can also be established by means of in situ testing which is more common in case of non-cohesive soils. Resonant column and torsional shear apparatus serve better for cohesive soils and give more consistent results then for loose soils (results much more scattered). In case of cohesionless materials *in situ* tests like surface waves method are preferred.

Concluding, as shown in both figures only the narrow bands of frequencies should be used to determine values of Rayleigh coefficients since in full spectrum they differ significantly.

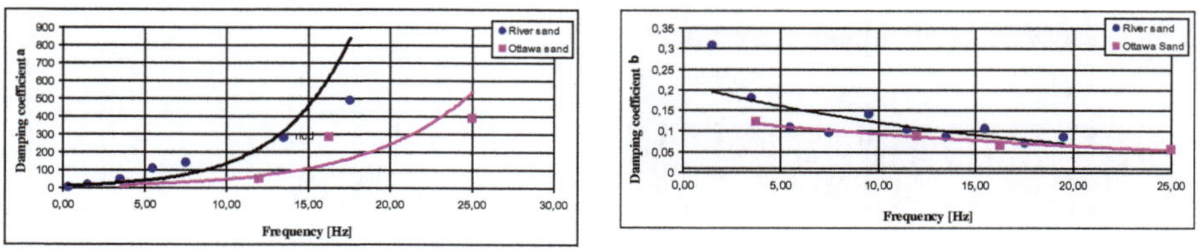

Fig. 2. Rayleigh damping coefficients vs. frequency for two types of sands.

2 Numerical Modeling

Numerical analysis (Finite Element Method here) is conducted for two selected problems and calculated in *Plaxis*. In the first case the underground structure with the horizontal impulse applied to the subsoil left boundary is considered. The second model describes vibro-isolation of the foundation of the generator implemented to protect the neighboring building. In both problems the wave propagation in porous medium and as a consequence the proper choice of Rayleigh coefficients takes major part. In purpose of drawing any decisive conclusions the physical values variation in time and space are investigated (mostly accelerations and velocities).

2.1 Underground Structure

Problem Definition

A simple rectangular (2,5 by 7,0 m) structure is placed under the surface at the depth of 5,5 m below the ground level. It is a model of a private family shelter capable of sustaining their lives in case of nuclear attack. To simulate the real threat the overpressure coming from the blast is apply to the model left boundary in the form of impulse load (lasting only 2 ms). Blast overpressure of 1380 kPa (200 psi) as being extremely destructive is assumed here and than convert to required input loading. As the magnitude of the explosives causes the crater of almost 90 m in depth the load is presumed to be linear and affecting the entire left side of the model. To the depth of 8,0 m lithology is composed of thin layers of cohesive soils (siCl, Cl/Si) followed by the deposit of sand and gravel fully saturated.

In theory the impulse loading should generate the full spectrum of frequencies, thus the calculations without damping are performed first in order to conduct Fourier transforms of the response in some representative points. The dominant frequency is established for vertical accelerations in *point H*, which is on the shelter most exposed wall and its value reaches 9,46 Hz. Fourier transformation and the rest of representative points are shown on Fig. 3. Then based on the preliminary analysis previously mentioned the proper values of *alpha* and *beta* for each soil layer are selected (see Fig. 1 and Fig. 2) and the model is recalculated.

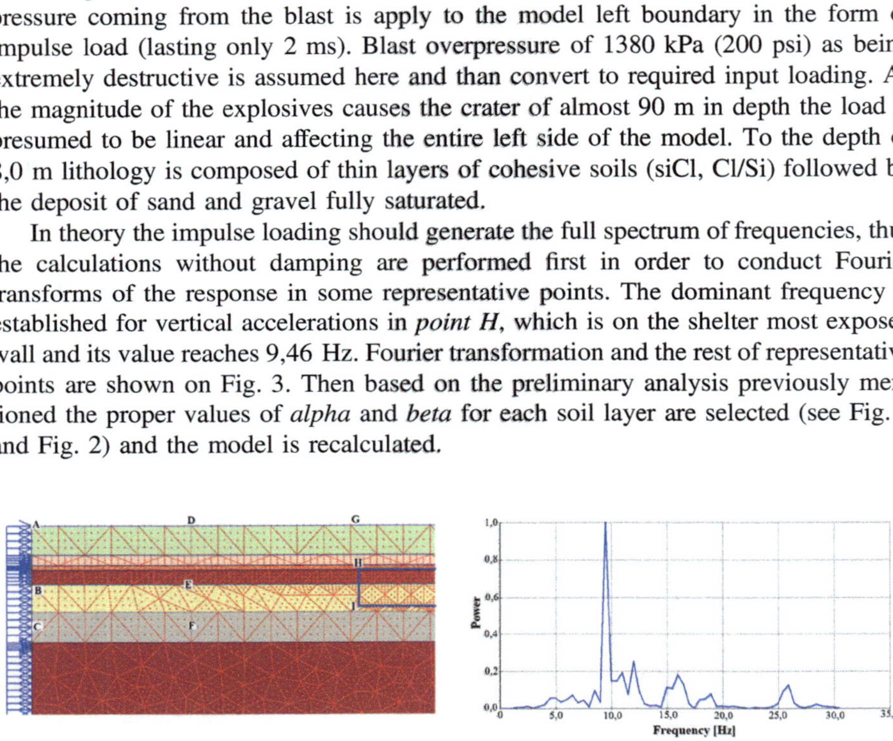

(a) (b)

Fig. 3. Representative points (a) and the Fourier transform of the response in point H (b).

Discussion of the Results

Without any damping shockwave resulting from the explosion reaches the structure of the shelter after 0,2 s (see Fig. 4) and causes the vibration similar to those encountered during earthquakes (compare Fig. 5a). All physical properties show the oscillating tendency and do not subside even after 2 s. After applying extreme values of Rayleigh coefficients, that is for the dominant frequency of 9,46 Hz they diminish significantly

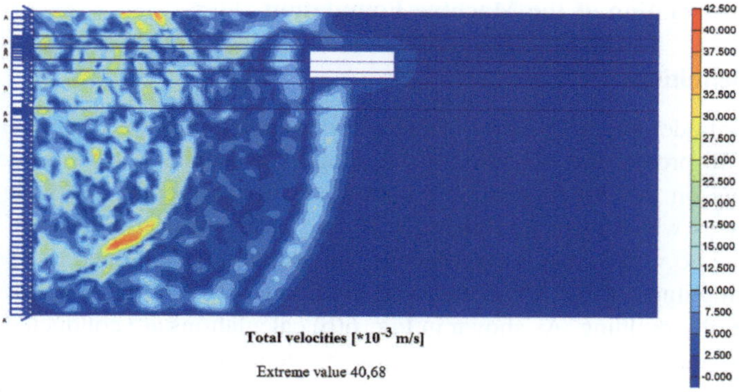

Fig. 4. Shockwave traveling thru subsoil (no damping implemented).

(some even completely) but the passage of the shockwave is finally visible on time curves (Fig. 5b). Surprisingly the blast overpressure reaches the structure much faster - in about 0,14 s. Therefore no-damping and the extreme damping create an envelope inside which the most realistic solution can be found. One way to deal with the problem is to perform separate calculations with *alpha* and *beta* for each frequency that will appear in representative points (important frequencies with which the subsoil responds). The final and most correct response of the soil can be then achieved as a superposition. This iterative procedure is absolutely possible but not entirely practical. From the designing point of view at least one calculation (for coefficients of the most expected or damaging frequency) should be made to prove that the damping reduces physical values in time but does not affects them too seriously in space. In both approaches the sensible results can be obtain by iterative procedure although no specific recipe can be written.

Summing up in this numerical example the damping coefficients clear the signal of the shockwave (filter the noises) in time domain but falsely reduce displacements and their time derivatives in space.

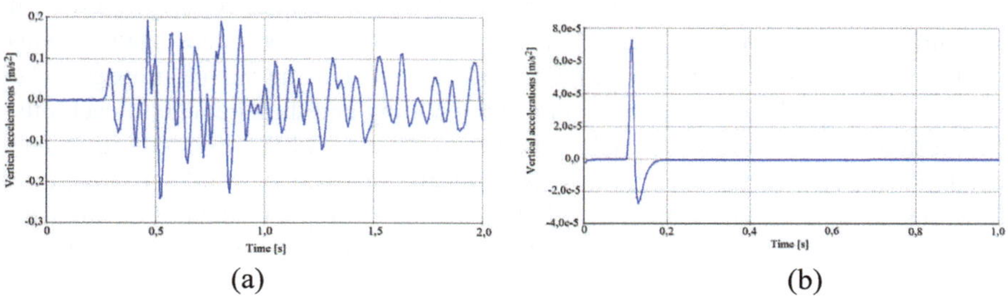

Fig. 5. Vertical accelerations in (point H): (a) before applying damping; (b) for a dominant frequency (extreme damping)

2.2 Vibro-Isolation of the Machine Foundation

Problem Definition

The second model describes vibro-isolation incorporated into the generator foundation in purpose to protect the neighboring building. The structure itself is modeled as a frame resting on the raft foundation - both made from concrete, but with quite large internal spaces which in turn influence it rigidity. Subsoil is composed of a cohesive soil therefore no water table is considered. The excitation is assumed to be a harmonic load with amplitude of 160 kN/m and working frequency of 10 Hz, located only 3,5 m from the nearby building. As shown in Fig. 6(b) calculations are conducted for 2 s and include working and disengaging phase of the generator. The representative points, presented in Fig. 6(a), are located at the structures and on the way of propagating wave and can easily be treated as the monitoring point.

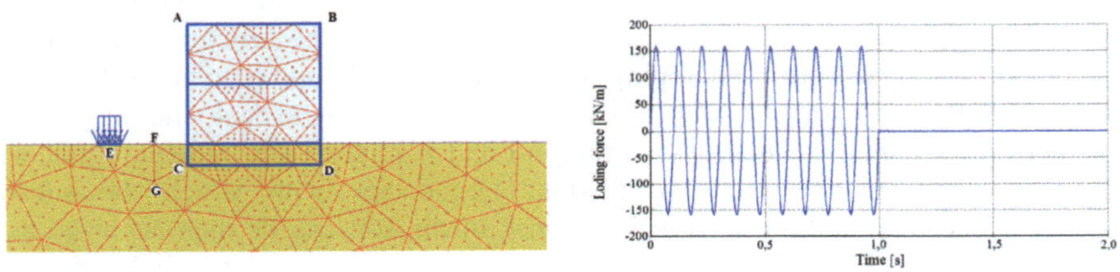

(a) (b)

Fig. 6. Representative points (a) and the input loading (b)

The main interest now is to reduce the vibrations produced by the generator and what fallows those that are reaching to the construction and its foundation.

Discussion of the Results

The problem itself seams to be totally different from the previous one mostly because of the source of excitations and the fact that the dominant frequency is known from the beginning. Now the loading located on the ground surface produces shallow waves which propagate in all directions. Figure 7 illustrates how the waves propagate from the source, bounce from the foundation of the building and then impede to each other. This behavior of the subsoil is only possible to observe while no damping is introduced. Similarly like in the previous example the absence of Rayleigh coefficients allows us to inspect the spatial distribution of physical parameters (Fig. 7), which clearly shows that the foundation is not only directly exposed to vibrations but also acts like a barrier from which the waves bounce back.

Unfortunately the lack of damping causes that physical parameters adopt unrealistic values in time, for example accelerations above 30 m/s^2 (see Fig. 8). In purpose to minimize this effect the Rayleigh coefficients are only implemented to the vibro-isolation. Figure 8(a) shows that without any isolation the building vibrations are

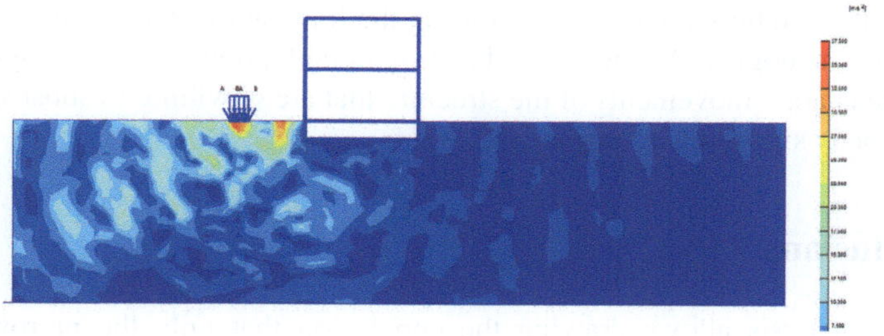

Fig. 7. Total accelerations after 1 s (no damping implemented).

almost random and its movements subside disbelievingly slowly after the generator is disengaged (points A and C are located on the most exposed wall of the structure).

But after the mat is inserted, the building starts to react much more realistically - the response is analogical to the excitation and finishes soon after the end of the input loading (Fig. 8b). The damping mat is modeled as a thin elastic plate with quite large α and β coefficients established on the base of product specifications. The vibro-isolation is placed right beneath the generator foundation and its effectiveness is best visible on the charts presented in and Fig. 9. Total accelerations at the top of the building's wall (point A) drops down to the more realistic value of around 0,7 m/s^2 (Fig. 9a).

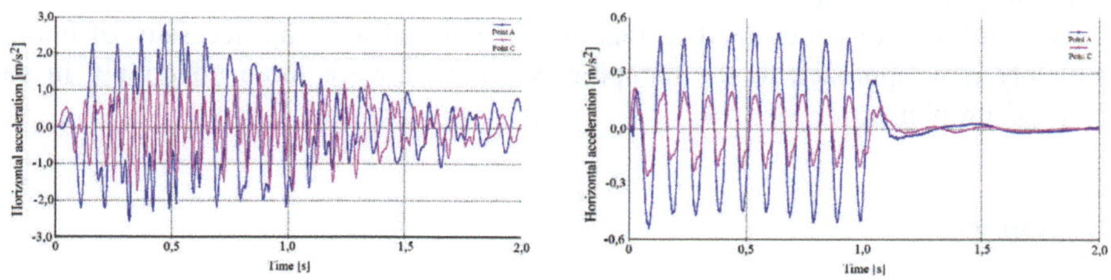

(a) (b)

Fig. 8. Horizontal accelerations of the building (point A and C): (a) without damping; (b) after the damping is implemented

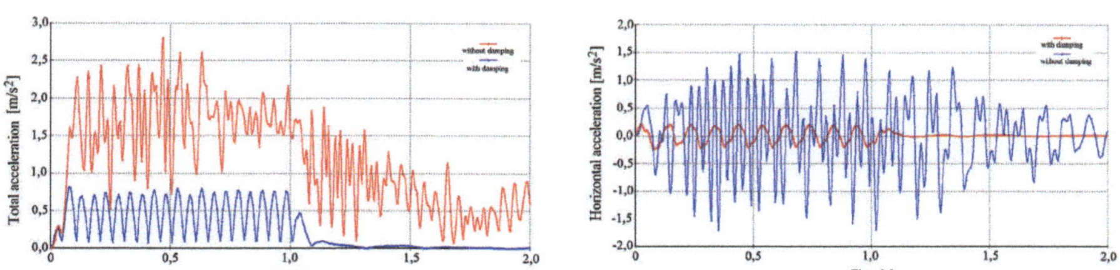

(a) (b)

Fig. 9. Physical values variation in time before and after applying Rayleigh damping: (a) at the top of the building (Point A); (b) at the foundation (Point C)

The reaction of the horizontal accelerations at the foundation level (point C) illustrated in Fig. 9(b) is analogical. Without any damping in both points A and C physical values demonstrate chaotic movements of the structure that are unwilling to subside even after the generator is switch off (compare Fig. 9).

3 Conclusions

Preliminary analysis allows drawing the conclusion that only the narrow bands of frequencies can be used to determine values of Rayleigh coefficients since in full spectrum they differ significantly.

Summing up the numerical calculation the general remark can be formulated: damping coefficients clarify the signal in time domain but reduce displacements and their time derivatives in space. Furthermore this phenomenon take place as soon any Rayleigh damping coefficients appear in the model - no mater where they are apply (to the subsoil as a whole or directly to the input loading). Hence to observe the overall behavior of the subsoil during excitation the use of damping is not recommended in any place. But on the other hand as the Rayleigh damping works as a filter removing the unnecessary noise for every physical value varying in time, it should be implemented to monitor those values properly in time domain.

Finding *alpha* and *beta* for which both wrong behaviors of the model will be excluded mean finding their proper values. It may be possible by conducting some iterative procedure mention above.

The considerations presented in this article exposed also the necessity of finding completely new approach to the damping which will work more effectively in soil dynamics' problems.

References

Mitchell J.K., Soga K.: Fundamentals of Soil Behavior. Wiley (2005). ISBN: 978-0-471-46302-3
Lin, M.-L., et al.: The effects of frequency on damping properties of sand. Soil Dyn. Earthq. Eng. Elsevier (1996). https://doi.org/10.1016/0267-7261(95)00045-3

Numerical Assessment of Slope Stability of Ain-Tinn Mila Province (Algeria)

Mohammed Bouatia(✉) and Rafik Demagh

University of BATNA2, Batna, Algeria
bouatia.mohammed@gmail.com

Abstract. The hydraulic project of Beni-Haroun has a particular importance in Algeria. It is part of an extensive program for the mobilization and transfer of surface water to address the strong regional hydrological inequalities. The difficult topography of Mila basin requires every time the assessment of the sliding risk of slopes which may contain complex system of water transportation from Beni-Haroun dam over long distances. In this paper, two and three-dimension numerical analysis are investigated to assess the safety factor of Ain-Tinn natural slope which already contain a buried pipeline with 800 mm diameter.

Keywords: Natural slopes · Instability · Numerical simulation

1 Introduction

In many part of the world, human lives and economic losses caused by landslides are very important [1]. The stability of natural slopes is a problem which interest both of geotechnical community as well researchers [2]. There is an increasing recognition of the need for assessment of landslide susceptibility and hazard and for landslide risk management [3].

The hydraulic project of the Beni Haroun dam is particularly important in Algeria. It is part of a vast water supply program designed to compensate the hydrological inequality between the different regions of Algeria by building a pipeline system.

But the difficult topography of the Mila region requires the installation of a complex technical system for the transfer of water from the Beni-Haroun dam (1 billion-m^3 capacity) over long distances. Giant pumping station, 3 control dams and over 600 km of pipelines [4].

Ain Tinn is a natural slope in Mila province (Algeria), this slope contains a water transport pipe of the dam of Beni-Haroun. This water transport system was commissioned in 2006 and after some years, several water leak points were noticed on the corridor of the pipelines and more exactly in the site of Ain Tinn, which disturbs the program of drinking water supply for citizens of the city of Mila and 12 other municipalities in the province of Mila.

Several hypotheses have been proposed by the Directorate of Water Resources (DRE) to know the causes of these water leaks which could cost more than 200 MDA to the hydraulic sector for the repair of a single break in a pipe of this size [5]. After an

H. Shehata and C. S. Desai (Eds.): GeoMEast 2018, SUCI, pp. 133–143, 2019.
https://doi.org/10.1007/978-3-030-01926-6_10

investigation with metal and geotechnical experts the slope instability is considered the main cause of rupture of these pipelines.

For this reasons, a more detailed investigation (field of research) appears to us important. This paper is a contribution to the assessment of landslide risk by processing geotechnical report data, preparation of the geometric and geotechnical slope model using CIVIL3D, FLAC, and FLAC3D softwares. Then a calculation of the security factor in 2D and in 3D is carried out using the strength reduction method implemented in the ITASCA software's FLAC and FLAC3D

2 Geographical and Geological Context

Mila Province is located in the North-East of Algeria. It is bordered north by Jijel and Skikda Provinces, south by Batna Province, West by Sétif Province and Constantine Province at East as shown in Fig. 1. It covers an area of 3,550 km^2 composed by five communes: Ferdjioua, Chelghoum Laid, Grarem Gouga, Oued Naja, Rouached and 32 municipalities.

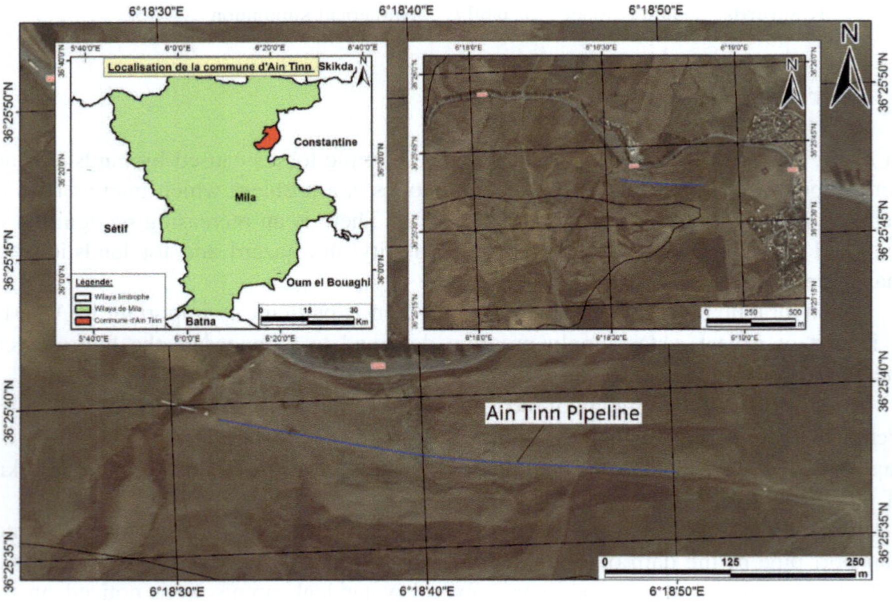

Fig. 1. Ain-Tinn slope localization

Ain-Tinn slope is located at 36°25′37″ North and 6°18′40″ East Fig. 1. His elevation varies from 610 to 686. Based on the in situ observations and the geological investigations, the slope has a gradient of 10–20°, with cracks and deformations

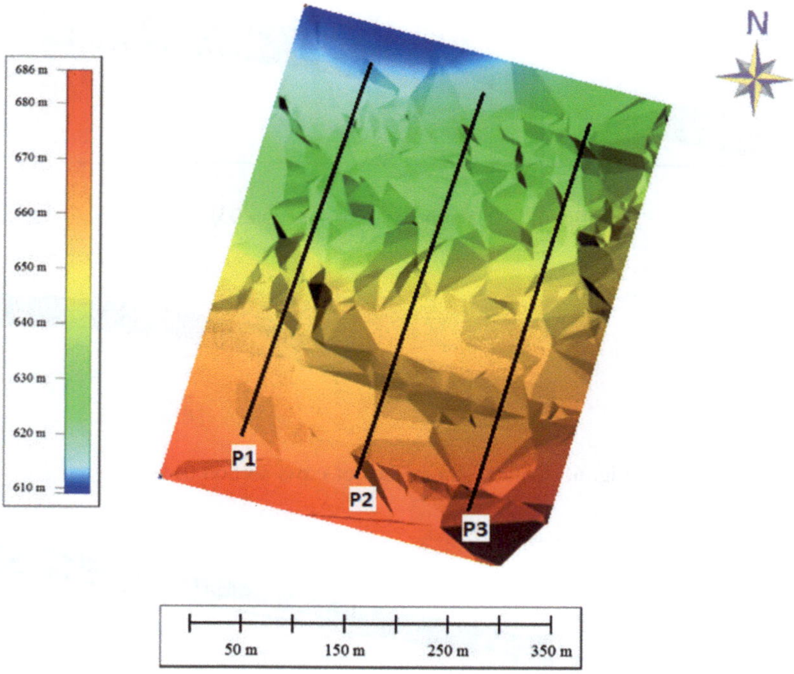

Fig. 2. DEM of Ain-Tinn slope with cross section demonstration

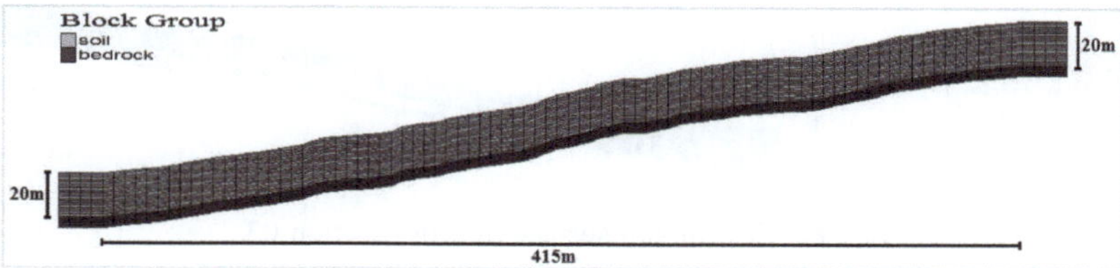

Fig. 3. 2D geometry (cross section 01)

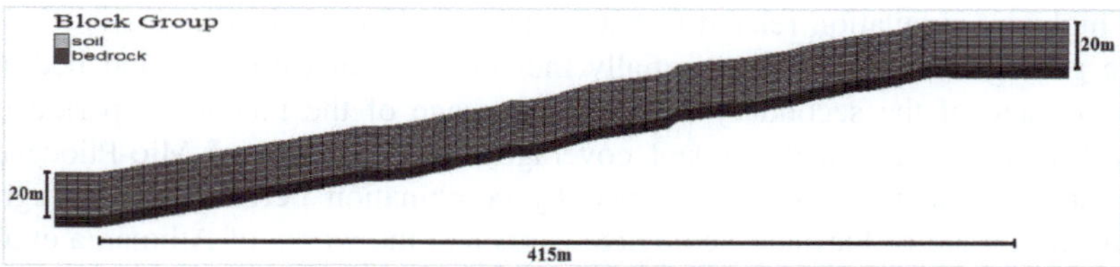

Fig. 4. 2D geometry (cross section 02)

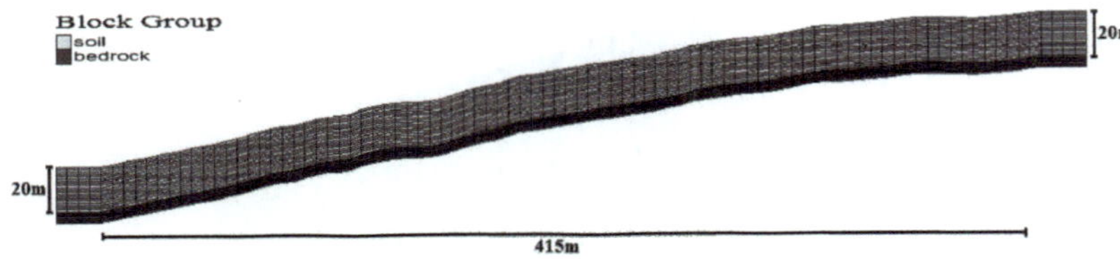

Fig. 5. 2D geometry (cross section 03)

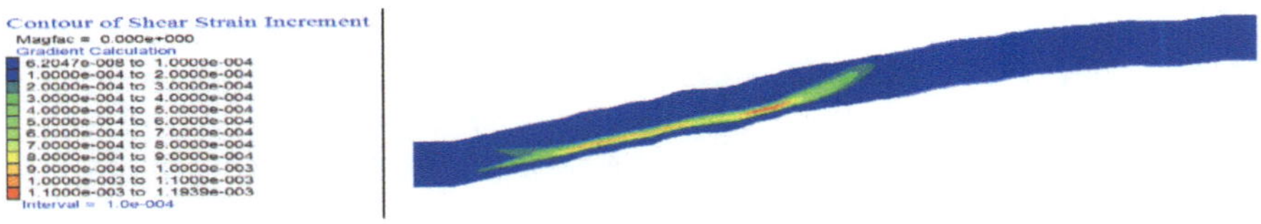

Fig. 6. Critical slip surface of cross section 01

Fig. 7. Critical slip surface of cross section 02

Fig. 8. Critical slip surface of cross section 03

appeared in the middle and crest of the slope, and concentrated in the shallow layer. For this reasons, specialists concluded that this slope had high potential to fail in the next rain season and endanger the water transfer pipeline on the middle of slope and break the 79 highway circulation related Constantine and Mila provinces.

The geological formations essentially include a set of carbonate and bedrock of Cretaceous age of the secondary Ere to Eocene age of the Paleogene period of the tertiary Ere. In surface, a discordant coverage which consists of Mio-Pliocene and Quaternary continental deposits obtained by combination between the stratigraphic simplification scale of Parriaux geological cards and the works of Athmania et al. [6].

These Mio-Pliocene formations cover the major part of Mila's basin. They are mainly represented by detritus deposits (conglomerates, gravel, sand, clays, evaporate deposits (gypsum, anhydrite and rock salt) and lagoon lime stones [6].

Geotechnical data presented in Table 1, was obtained from numerous reports produced by LNHC (Laboratoire National de l'Habitat et de la Construction) geotechnical investigations, and derived from both in-situ and laboratory tests. The last geotechnical report indicates the existence of a uniform gradient of the cohesion with the depth which equal to 1 kPa/m, the initial cohesion at surface is equal to 20 kPa. During the Standard Penetration Test (SPT), the refusal was reached at 20 m depth [7].

Table 1. Slope mechanical parameters [7]

Layer	Weight (kN/m^3)	Initial cohesion (kPa)	Friction angle (degree)	Tension (kPa)	Bulk (GPa)	Shear (GPa)
Soil	20.0	20	12	5	2.80	0.80
Rock	21.7	100	40	50	25	15

The interface between soil and bedrock is considered as parallel to the ground surface of the slope and having constant depth of 20 m.

3 SRM Approach and Mohr Coulomb Failure Criterion

The soil mechanics definition of the safety factor is the ratio of the available shear strength to the minimum shear strength needed for equilibrium. In other terms, it is a ratio of the availability to the need, considering the shear strength of soil.

In this paper the strength reduction method (SRM) founded in Mohr-Coulomb failure criterion as described in [8] work is used to obtain the factor of safety. The finite element or finite difference methods use a different criterion of failure which gives a more real simulation of soil behavior and deformation at first and second to consider the effects of deformation development in soil massif characteristics. The shear strength of soil which needed for design is given as:

$$\tau_f = c + \sigma_n tan\varphi \quad (1)$$

Where τ_f is the shear strength of the soil, σ_n is the applied normal stress, c and φ are the strength parameters of the soil (cohesion and internal friction respectively).

The reduced shear strength τ_m along the failure surface is expressed as follows:

$$\tau_m = \tau_f / f \quad (2)$$

$$\tau_f = c/f + (\sigma_n tan\varphi)/f \quad (3)$$

The value of f is adjusted until the slope fails, where the ultimate f is the factor of safety obtained by 1e-5 value of maximum unbalanced force ratio [9].

4 Preparation and Building of the Geometry

Ain Tinn is a natural slope modified by human activities (construction of roads and pipelines), which makes its geometry complex.

The Autodesk-CIVIL3D engineering design software is used to realize a digital elevation model (DEM) from the elevation contours relative to the project site implemented by a topographic brigade of the Hydro Project Est Society (HPE).

In this section, the objective is to illustrate the use of the FISH language to create a surface model that corresponds to the elevation contours. The plan dimensions of the primitive elements are 5 × 5 m. The problem is in the determination of the elevations of the vertices of the primitive of construction of the model which is the wedge in this case there.

Several sections are taken with constant y coordinates. A table is created for each section, these tables will contain pairs (x, z) for each point of intersection with an elevation contour. These tables are grouped in an ASCII file which will be read by the program developed using FISH language to interpolate the value of Z for a given X first, and secondly the generation of the model based on a series of commands by assembling wedge elements as shown in Fig. 9 when adopting all profiles in the y direction for a 3D model, and assembling brick elements as shown in Figs. 11–13 when selecting cross sections for a 2D Modeling.

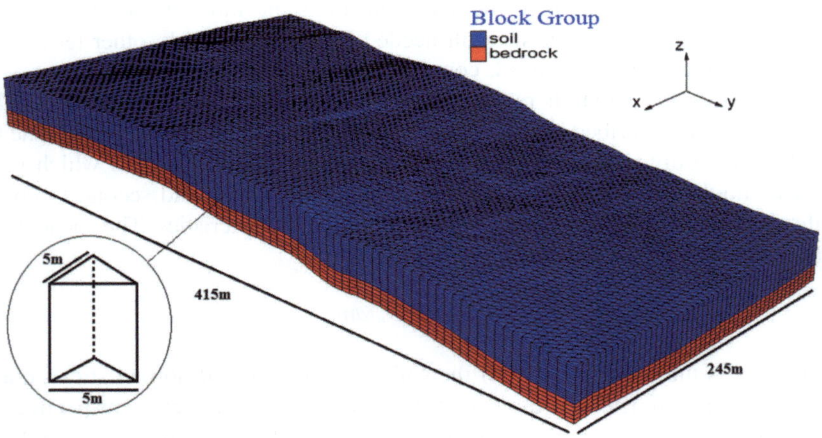

Fig. 9. 3D numerical model of Ain-Tinn slope.

5 2D Numerical Modeling and Analysis

5.1 2D Modeling

In order to simplify simulations and have quick results by choosing the most vulnerable cross sections, which requires certain conditions to be applicable (symmetry and

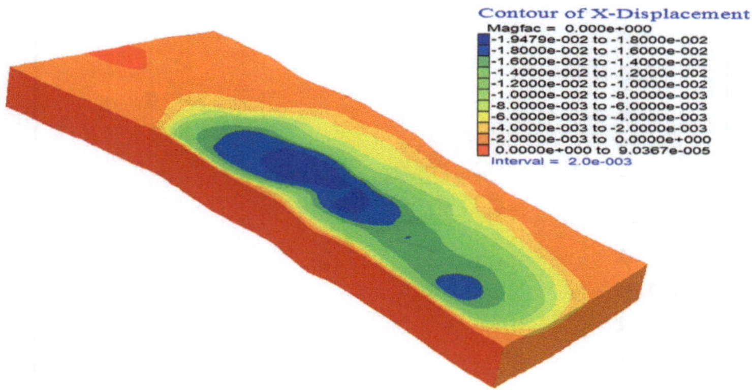

Fig. 10. 3D x displacement simulation results

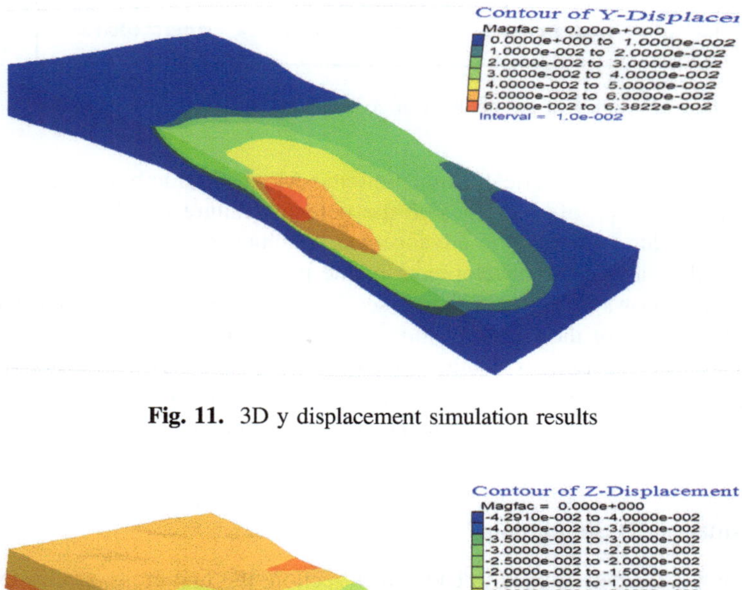

Fig. 11. 3D y displacement simulation results

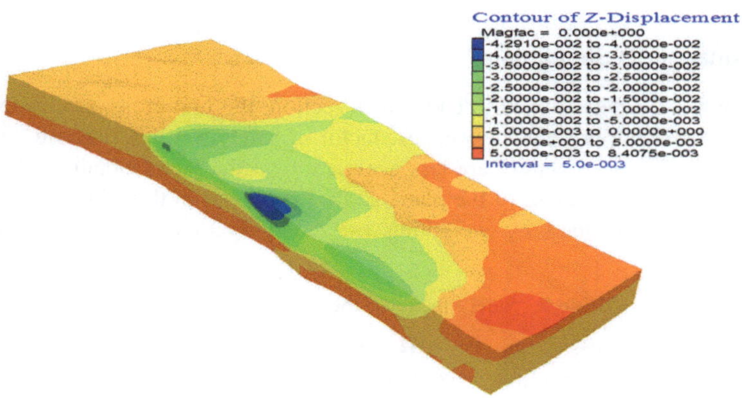

Fig. 12. 3D z displacement simulation results

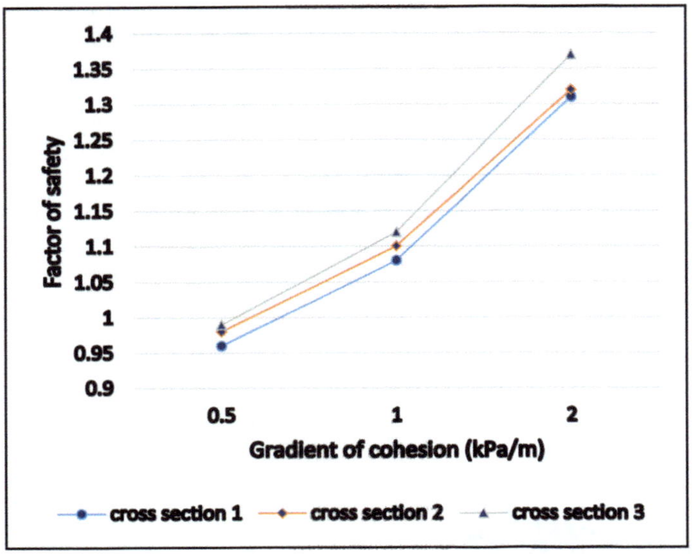

Fig. 13. Effect of the gradient of cohesion on the 2D Factor of Safety

homogeneity). In 2D numerical simulation the Mohr Coulomb failure criterion is adopted, and only the safety factor and the potential failure line for some particular cross section which is selected in advance can be obtained [8].

As described above the development of the program of geometry building using FISH language allows to choose the cross sections in any position, and to install a non-uniform distribution of the characteristics according to the depth (i.e. the cohesion).

The three cross sections which are subjected to this analysis are located in middle (P2: Cross-Section-02), west (P1: Cross-Section-01) and east (P3: Cross-Section-03) as shown in Fig. 2. Natural slopes and 2D grid geometry are shown in Figs. 3, 4, 5 respectively.

5.2 Results of 2D Simulations

Table 2 show the safety factors in each cross section, the first cross section located in the west of the slope present the weakest factor of safety, on this purpose we should note that these factors of safety are less accuracy than the factor obtaining by the 3D numerical simulation because of the plane-strain hypothesis in this 2D simulation. The average value is obtained by weighting the safety factors by the surface of the

Table 2. Safety factor of the three cross sections

Cross section	**01**	**02**	**03**	Weighted average
Safety factor	**1.08**	**1.10**	**1.12**	**1.10**

respective soil masses in displacement [3]. These values are still low, which requires further investigation in 3D numerical analysis to appreciate the global stability of the massif and in order to obtain more realistic safety factor values.

Figures 6, 7, 8 shows the critical slip surface of the different cross sections drawing effectively by the contour of displacement according to the y direction with a ratio of maximum unbalanced force equal to 1e-5 [9].

It is easy to remark the existence of different form and scale of the critical slip surface. These figures also show the existence of a shear plane close to the soil-bedrock interface. The parallel shape of the sliding surfaces with the natural surface also means that the landslide of Ain-Tinn slope is a translational landslide.

6 3D Numerical Modeling and Analysis

6.1 3D Modeling

For complex and heterogeneous slopes, 2D stability analysis is inappropriate and leads to an underestimate of the safety factors with inaccurate results [8]. Ain-Tinn slope is a heterogeneous slope because of the complex topography and morphology of the site at first and because of the different shapes of the critical slip surfaces allowed in 2D simulation secondly. Using the Mohr coulomb failure criterion, 3D numerical analysis was carried out for more accuracy results but it will be take a significant resolution time because of the high number of nodes and zones comparing with the 2D model.

Using the program of geometry construction developed in this study, The 3D model in finite difference is constructed by assembling a wedge element of 5 m × 5 m matrix in (x, y) direction and variable in z direction as shown in Fig. 9 to obtain 160.720 zones and 87.150 nodes as numerical size of the 3D model, divided in two layers. The first denser mesh layer is the soil of 20 m thick, the second is the bedrock with less mesh density.

6.2 Results of 3D Simulation

Based on the Mohr coulomb yield criterion with the geotechnical parameters listed in Table 1, The 3D safety factor calculated using FLAC3D strength reduction method is equal to 1.14, which is higher by 5.5% than the minimum 2D safety factor (i.e. 1.08), higher by 3.6% than the average 2D weighted factor of safety (i.e. 1.10). After the shear strength has been reduced by a coefficient of 1.14, the displacements in x, y and z directions and the slope state are shown Figs. 10–12.

The major displacements on Y direction are concentrated in the eastern side of slope (represented by cross Sect. 3). The major displacements on x direction are concentrated in the eastern side of slope (represented by cross Sect. 3).

7 Parametric Study

In order to study the effect of the cohesion gradient on the value of the safety factor, a parametric study was performed both in 2D and 3D, Figs. 13 and 14 respectively. As expected, in 2D investigation, the parametric study, shows that the safety factor varies from 0.96–0.99 to 1.3–1.37 depending on whether the cohesion gradient varies from 0.5 to 2 kPa/m depth. In the same time, in 3D investigation, the results show also the same trend and the safety factor varies from 1.01 to 1.39.

These results show the dependence of the safety factor with the variation of gradient of cohesion, which is not always known correctly. This affects the appropriate choice of the resistance parameters to be taken into account in order to more realistically estimate the safety factor. The calculation times remain nevertheless very penalizing in 3D simulations.

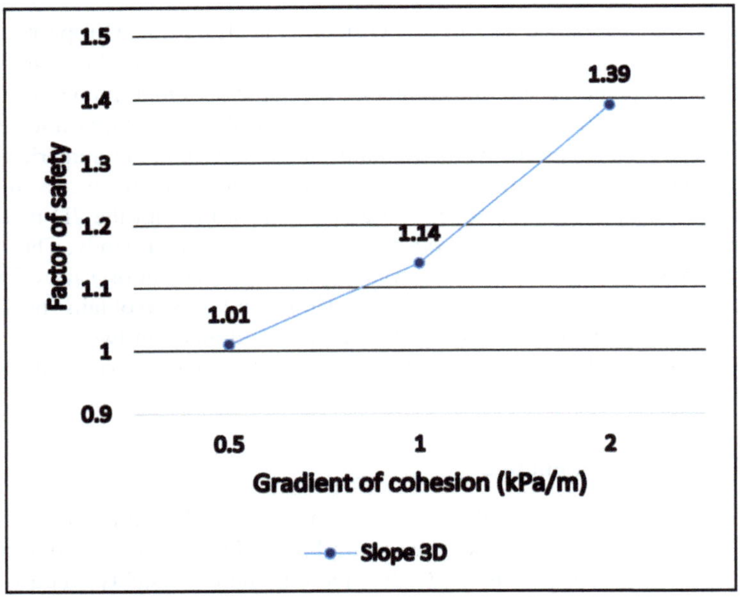

Fig. 14. Effect of the gradient of cohesion on the 3D Factor of Safety

8 Conclusion

In this paper, 2D and 3D numerical simulation of Ain-Tinn slope stability sliding susceptibility which contain a buried pipeline of water transfer was carried out. The potential of stability is appreciating in terms of safety factor and displacements field. The simulations in short term (undrained conditions) were carried out using the finite difference code FLAC and FLAC3D. This investigate study is beginning by preparing the 2D in plane strain model and 3D model (global stability).

The application to the case of Ain-Tinn gives very weak safety factors under natural conditions. These same safety factors would be revised downwards in the case of rapid saturation of the massif (natural saturation due to high rainfall or an accidental break of the water pipeline for example). The results of the 3D simulations show that the maximum displacements are concentrated in the middle of the slope, so the risk is all the greater as the buried pipe is close to the middle of the natural slope. In anyway, these results are comforted by the 2D simulations also.

This study allowed us to confirm that the safety factors calculated using 2D methods are more conservative than those estimated using the 3D model, more realistic and taking into account the overall slope of the massif with all its topography. Moreover, a poorly justified choice of any cross-section can lead us to make decisions that are too hasty and risky, as is the case with the cross-Section 01 if it had been taken alone as reference to estimate the stability of the slope of Ain-Tinn.

Finally, to study the effect of the gradient of the cohesion, a parametric study is carried out in 2D and in 3D by using a program developed by language FISH of Code FLAC3D, it allows us to initialize the cohesion according to the depth. The results of this study showed that this parameter has a very important effect on the safety factors of Ain Tinn slope.

Acknowledgement. The authors want to knowledge SETSétif, LNHC Batna, DRE Mila, ANBT Alger.

REFERENCES

1. Conforth, D.H.: Landslides in Practise, Metu Library. Wiley, New Jersey (2005)
2. Khemissa, M.:"Methodes d'analyse de la stabilite et techniques de stabilisation des pentes," presented at the Journées Nationales de Géotechnique et de Géologie de l'Ingénieur - JNGG' 2006 Lyon (France), (2006)
3. Chowdhury, R.: Geotechnical Slope Analysis. Taylor & Francis Group, CRC Press/Belkema, London, UK (2010)
4. Bouatia, M., Demagh, R., Khannouf, A.: Assessment of Slope Stability of Ain-Tinn Mila Province (Algeria), Presented at the International Simposium on Construction Management and Civil Engineering - ISCMCE 2017 Skikda, Algeria (2017)
5. Bouatia, M., Demagh, R.: 3D Numerical Analysis of Slope Stability (Case of Ain Tinn Mila Pipeline), Presented at the 1er Siminaire National Geologie, Geotechnique, Georessources 1er 3G' 2017. Jijel, Algeria (2017)
6. Athmania, D., Benaissa, A., Hammadi, A., Bouassida, M.: Clay and Marl formation susceptibility in Mila province, Algeria. Geotech. Geol. Eng. **28**(6), 805–813 (2010)
7. LNHC-Batna, "Rapport d'étude de sol : Etude de confortement ou la déviation de la conduite AEP du couloir 1 de a station de la station de traitement Ain Tinn_Mila," LNHC, ALGERIA, géotechnique (June 2016)
8. Shen, H., Klapperich, H., Abbas, S.M., Ibrahim, A.: Slope stability analysis based on the integration of GIS and numerical simulation. Autom. Constr. **26**, 46–53 (2012)
9. ITASCA 2003, FLAC3D version 3.1 Fast Lagrangian Analysis of Continua in Three Dimensions. Itasca Consulting Group, Inc, Minneapolis, Minnesota, USA (2003)

Stochastic Modeling of the Spatial Variability of Soil

Rubi Chakraborty[1] and Arindam Dey[2(✉)]

[1] Department of Civil Engineering, National Institute of Technology, Meghalaya 793003, India
rubi.rc.nit@gmail.com

[2] Department of Civil Engineering, Indian Institute of Technology Guwahati, Guwahati 781039, India
arindamdeyiitg16@gmail.com

Abstract. Disparate sources of uncertainties contribute to the variability in geotechnical properties. Inherent variability in soil property, error in measurement and uncertainty in transformation models are the three major sources of geotechnical uncertainties. This paper mainly focusses on inherent soil variability; while the rest are beyond the scope. The inherent variability in soil originates mainly from the natural geological phenomena that result in the origination and the continual modification of the in situ or residual soil mass. Inherent soil variability is generally modelled as a random field. Simulating geotechnical variability as random process incorporates significant statistical results to be inferred from field test data and helps in including such variability in reliability analysis of geotechnical design. In this study, a stochastic model for inherent soil variability in terms of SPT-*N* value is developed for IIT Guwahati campus (26.1903° N, 91.6920° E) situated in Guwahati city of state Assam in India. Total 146 borehole SPT-*N* profile of the campus are used for developing the model. SPT-*N* value is observed to have an increasing trend with depth i.e. the data is nonstationary. In this model, the original data is first detrended and the trend function can be treated as deterministic in nature. The trend function is then evaluated using linear regression; a linear model is found to be good enough for present data. The residual part (error) is modelled as a one-dimensional random field. The squared exponential correlation structure is found to be best fitted in this case. The scale of fluctuation of the random field characterizing the error has been estimated from the fitted autocorrelation structure. The errors can be simulated easily once the correlation structure is known and added to the deterministic part to generate desire number of *N* profiles for further geotechnical analyses of the site.

1 Introduction

The variability in any geotechnical property originates due to different sources of uncertainties, as shown in Fig. 1. The three major sources of uncertainty in any geotechnical property are inherent soil variability, error in measurement and uncertainty related to transformation models. The natural geological phenomenon of producing soil and its continuous modification results in the inherent variability in any soil

H. Shehata and C. S. Desai (Eds.): GeoMEast 2018, SUCI, pp. 144–155, 2019.
https://doi.org/10.1007/978-3-030-01926-6_11

in situ property. The second source i.e., the measurement error originates due to equipment, operation and other testing errors. Generally, these two sources are collectively named as data scatter (Kulhawy et al. 1991). Statistical uncertainty or sampling error resulting from limited available information may influence in situ measurements, which can be minimized by taking more samples (Kulhawy et al. 1991). The third source of uncertainty is due to empirical or other correlation models used for transformation of field or laboratory measurements into design soil properties. The relative contribution of these uncertainty sources in the design soil property solely depends on the site condition, equipment used for laboratory or field tests conducted, error in operation and accuracy of the correlation model in use.

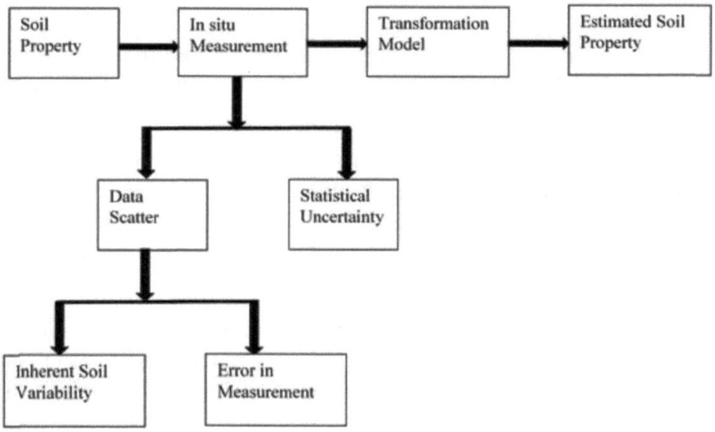

Fig. 1. Sources of geotechnical uncertainty in soil

This paper mainly focusses on inherent soil variability; while the rest are beyond the scope. Soil is a complex engineering material originated by various geological, environmental and physical-chemical processes. Moreover, these processes are modifying the soil in situ continuously. Due to these continuous natural processes, all in situ soil properties vary significantly, both in vertical and horizontal direction. In real world, available data are limited and analytical capacity is finite. The spatial variation of soil properties is modelled using mathematical random processes not because soil properties are random, but because the knowledge about the variation in soil in situ properties is limited.

This spatial variation in any soil property can be decomposed into a mean trend function $t(z)$ and a fluctuating component $w(z)$ as follows (Fig. 2):

$$\xi(z) = t(z) + w(z) \tag{1}$$

in which ξ is the in situ soil property and z is the depth (DeGroot and Baecher 1993; Phoon and Kulhawy 1999). The inherent soil variability is represented by the fluctuating component $w(z)$ defined in Eq. (1). The traditional way to quantify inherent

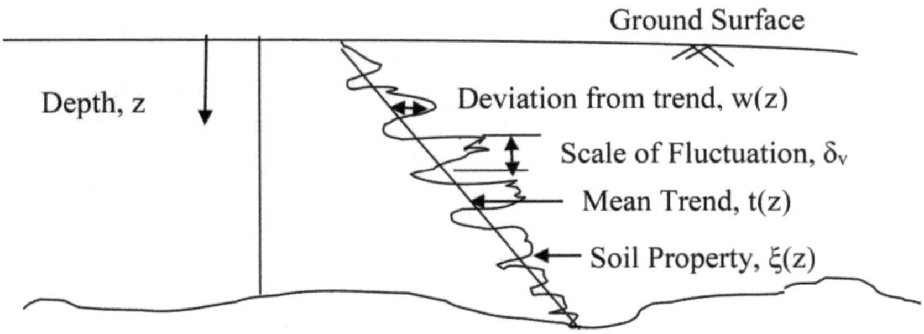

Fig. 2. Inherent variability in soil property

variability in soil is to model *w*(z) as a homogeneous random field (Vanmarcke 1983). The function related to fluctuating component *w*(z) will be statistically homogeneous if

(i) the mean and variance are constant with depth; and
(ii) the correlation between the soil property at two different points is independent of their absolute positions and depends only on the distance of separation. This correlation represents linear dependence between two random values varying between +1 and −1. A correlation of +1 or −1 means a perfect linear relationship between the two random quantities.

Uzielli et al. 2005 characterized normalized cone tip resistance and friction ratio (FR) rigorously using a finite-scale weakly stationary random field model. Raghukanth and Dash (2008) developed a stochastic model of SPT data for evaluation of probability of liquefaction in Guwahati City. Stuedlein et al. (2012) presented the geotechnical characterization and random field modeling of Beaumont clay at a field test site near Houston, Texas. In the present work, a stochastic model for inherent spatial variability of soil in terms of SPT-*N* value is developed considering IIT Guwahati campus as the case study site. The original data is first detrended and the trend function is evaluated using linear regression. The residual part (or error) is modelled as a one-dimensional homogeneous random field with the aid of squared exponential correlation structure.

2 Present Study

In Standard Penetration Test (SPT), a standard split-spoon sampler is driven into the borehole by the means of repeated blow of a standard hammer weighing 63.5 kg falling through a standard height of 750 mm at the rate of 30 blows per minute and the number of blows (*N*) required for a standard penetration of 300 mm is recorded (IS-2131-2002). Soil with low shear strength will give less value of *N*, whereas for a soil with high shear strength the *N* value will be comparatively higher. Spatial variability in SPT-*N* value is highly erratic in nature because of the inherent variability in soil property from one place to another resulting from various geological or physical-chemical process and continuous change in overburden stress. The measurement error also plays a vital role in case of SPT. Also, the recorded spatial variation of *N* values is not continuous as SPT value is recorded at certain discrete depths in the borehole. Moreover, the recorded *N* value represents a local average over some representative volume of the soil in the borehole. Therefore, modelling of uncertainties in SPT-*N* value is highly complex task and need to be modelled as a random field. Moreover, SPT-*N* value recorded shows a strong non-stationary behaviour, i.e. an increasing trend with increase in depth. Therefore, in this work, uncertainty in SPT-*N* value is characterized as a stochastic model by taking IIT Guwahati campus (situated in the state of Assam, North East part of India at 26.1903^{o} N, 91.6920^{o} E) as a case study site. The spatial variation is decomposed into non-stationary part (increasing mean trend) and a fluctuating component (error after detrending the data). The non-stationarity of the SPT-*N* data is simulated through linear regression in this work. The fluctuating component i.e. error is characterised as a one-dimensional homogeneous random field simulated using auto correlation function (ACF). Different theoretical ACFs are computed for this case study and squared exponential correlation structure is found to be best fitted in this particular case. With the use of this stochastic model derived in this study, a desire set of SPT-*N* data can be generated for further geotechnical analyses.

To develop this stochastic model, the SPT-*N* values from 146 borehole profiles distributed over different parts of IIT Guwahati campus are used. Figure 3 presents the IITG campus map showing the different borehole locations. These data are collected from different construction companies associated with the constructions in IIT Guwahati campus. The SPT-*N* data are recorded at an interval of 1.5 m for a depth of 30 m, in most of the locations. Figure 4 shows a typical borehole profile in campus. Figures 5 and 6 show the variation of *N* values and trend of mean values of *N* with depth respectively. It is observed that the SPT-*N* data shows an increasing trend with depth. Figure 7 shows the increasing trend of standard deviation of SPT *N* value obtained in field test.

2.1 Stochastic Modelling of SPT-*N* Value

Variation of SPT-*N* value with depth is highly erratic and also shows an increasing trend (Figs. 5 and 6). At all boreholes, this non-stationary behaviour in the SPT-*N* is observed (Fig. 5). Stationarity in the SPT-*N* can be achieved by different methods namely trend removal, variance transformation and differencing (Jaksa 1995). Stationarity in data can be defined as data having a constant mean and an autocovariance function depending only on separation distance rather than their absolute positions. The most commonly used method for achieving stationary fields is the removal of trends (Fenton 1999; Cafaro and Cherubini 2002; Akkaya and VanMarcke 2003). Therefore, the foremost aim is to remove this spatial increasing trend in the SPT-*N* data to acquire stationarity in SPT data. To achieve this, the SPT-*N* values can be expressed as summation of spatial mean trend (N_{avg}) and error part (w) as

$$N(z) = N_{avg}(z) + w(z) \tag{2}$$

where, z is the depth of soil in borehole.

Fig. 3. IIT Guwahati campus map showing the borehole locations marked by red star (source: http://www.iitg.ac.in/aa/pages/campusmap/index.php?hq)

In this model, the spatial trend is approximated by linear regression through linear (Fig. 8) as well as quadratic model (Fig. 9), as expressed in Eqs. (3) and (4) respectively.

$$N_{avg}(z) = 1.6508z + 0.8065 \tag{3}$$

$$N_{avg}(z) = 0.068z^2 - 0.0823z + 8.6053 \tag{4}$$

Description of strata	Depth
Filled up soil	2 m
Sandy clay grey in colour	4 m
Clay grey in colour	8 m
Sandy clay ash in colour	11 m
Coarse sand	13.5 m
Clay ash in colour	15 m
Coarse sand mixed with pebbles	16 m
Water table at 1.8 m below ground level	

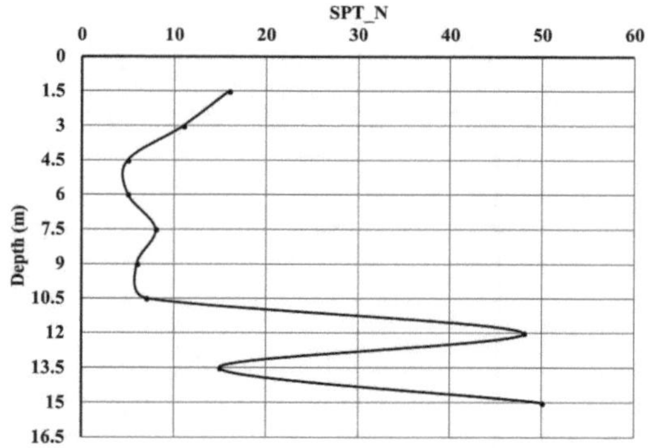

Fig. 4. A typical SPT-*N* profile in IIT Guwahati Campus

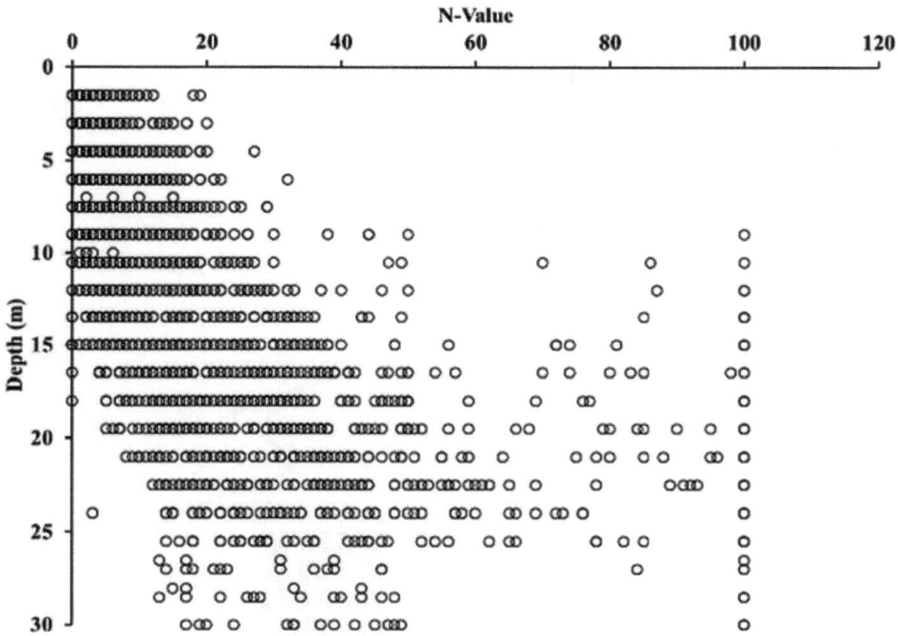

Fig. 5. Variation of SPT-*N* value with depth

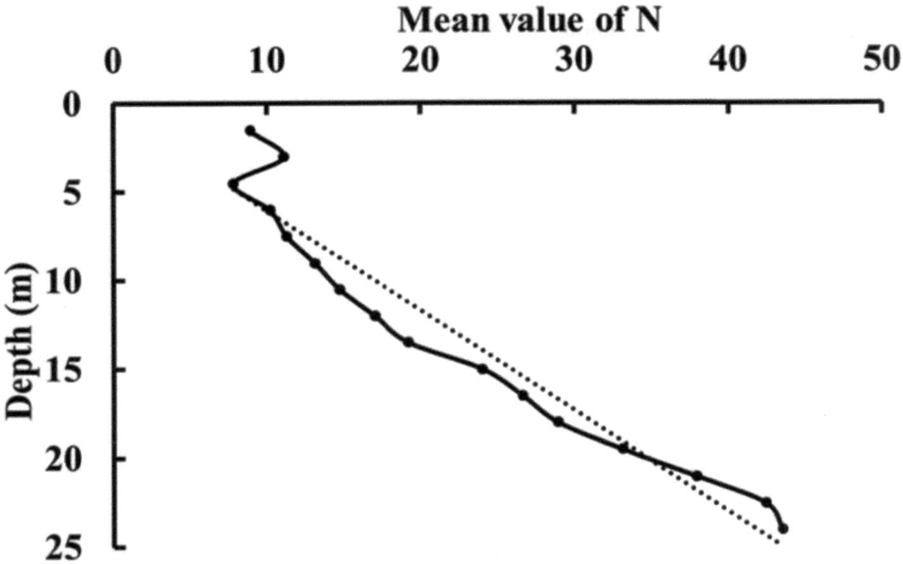

Fig. 6. Increasing trend of mean value of SPT-*N* with depth

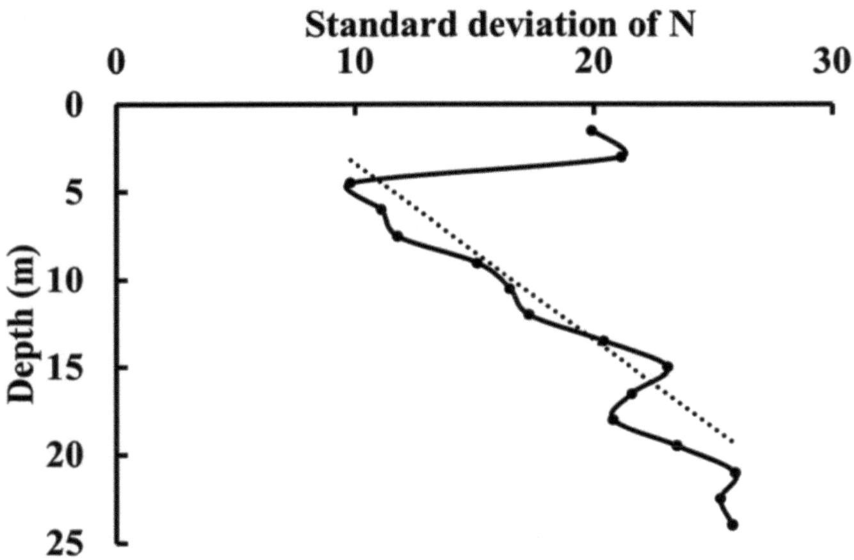

Fig. 7. Variation of standard deviation of SPT-*N* value with depth

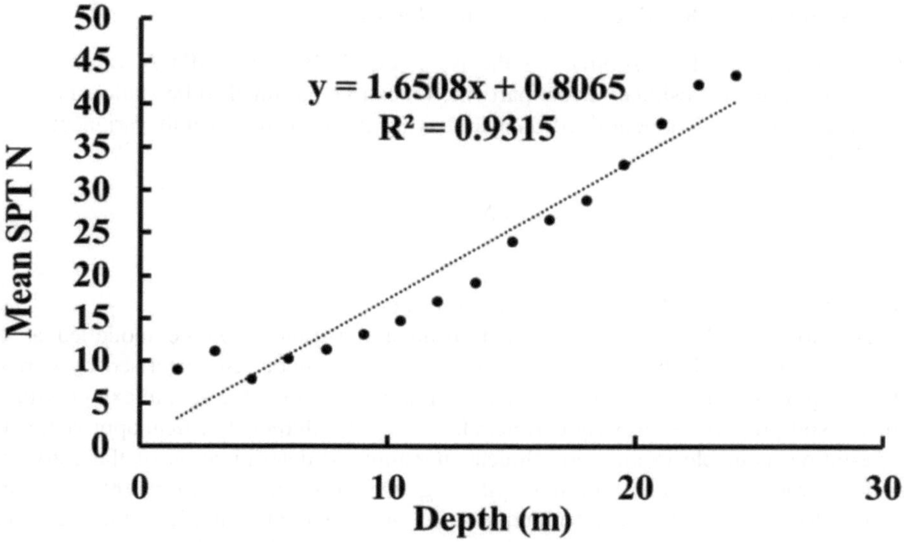

Fig. 8. Linear approximation of mean trend in SPT-*N* data

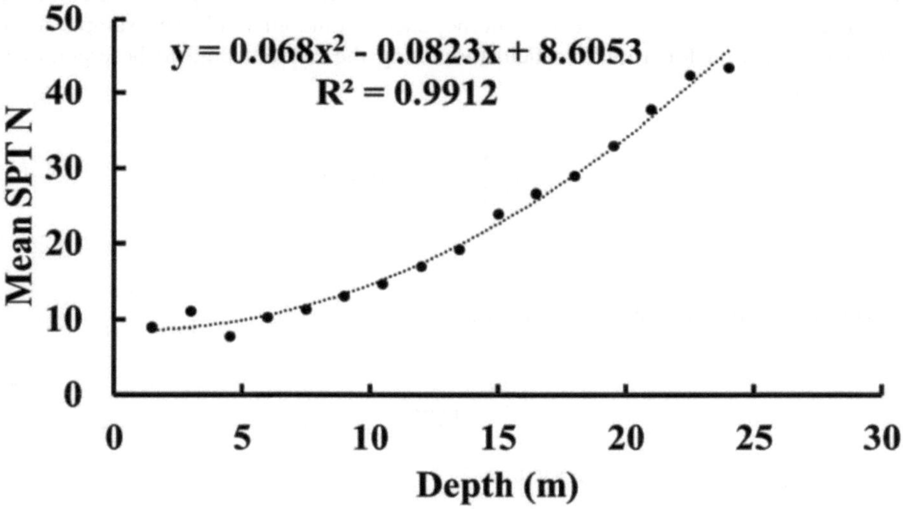

Fig. 9. Quadratic approximation of mean trend in SPT-*N* data

It is found that the linear model for N_{avg} is good enough for present data set, therefore for simplicity linear model is recommended for further analyses. Equation (3) is recommended to be used for detrending the data for further statistical analysis. Trend removal using linear regression is discussed in different literatures (e.g. Campanella et al. 1987; Popescu et al. 1998; Elkateb et al. 2003; Phoon et al. 2003; Uzielli 2004).

2.2 Modelling of Residuals After Trend Removal

Error w is estimated by subtracting the N_{avg} (Eq. 2) from the SPT-N value. After removing the trend, residual (error) part (i.e. w) can be assumed to be stationary. It is convenient to model standardized data $\bar{w}$, having zero mean and unit variance; estimated as

$$\bar{w}(z) = \frac{N(z) - N_{avg}(z)}{\sigma(z)} \tag{5}$$

where, $\sigma(z)$ is the standard deviation in the errors.

The normalized error with constant mean and variance can be modelled as a Gaussian normal field by means of an autocorrelation model characterised by correlation length (or scale of fluctuation). The correlation length is the spatial extent within which a soil property is strongly correlated. There are different statistical approach and curve-fitting available in the geotechnical literature for determination of the autocorrelation structure and correlation length (e.g. DeGroot and Baecher 1993; Fenton 1999). Most commonly used technique is to determine the sample autocorrelation function (ACF) and fitting a theoretical autocorrelation model (ACM) to the determined ACF. The correlation length can be estimated from the model parameter in the fitted ACM.

Now, an auto-correlation function is used to characterize the spatial variability at two different locations in space in a homogeneous random field. The sample autocorrelation function for a random field (assuming the random field to be ergodic) is expressed as:

$$\rho(\tau_j) = \frac{\bar{C}(\tau_j)}{\bar{C}(0)} = \frac{1}{\sigma_x^2}\left[\frac{1}{n}\sum_{i=1}^{n}(X_i - \mu_X)(X_{i+j-1} - \mu_X)\right] \tag{6}$$

$$\mathrm{j} = 1, 2, \ldots\ldots, \mathrm{n}$$

The sample average of x_i is computed as,

$$\mu_X = \frac{1}{n}\sum_{i=1}^{n} X_i \tag{7}$$

and the sample variance is expressed as,

$$\sigma_x^2 = \frac{1}{n}\sum_{i=1}^{n}(X_i - \mu_X)^2 \tag{8}$$

The estimated sample ACF of the standardized error of all the borehole SPT-N profile is shown in Fig. 10. Many theoretical 1-D correlation models (ACM) are available in the geotechnical literature. Table 1 shows three such correlation models commonly used in literature namely the exponential, squared exponential and second-order Markov field.

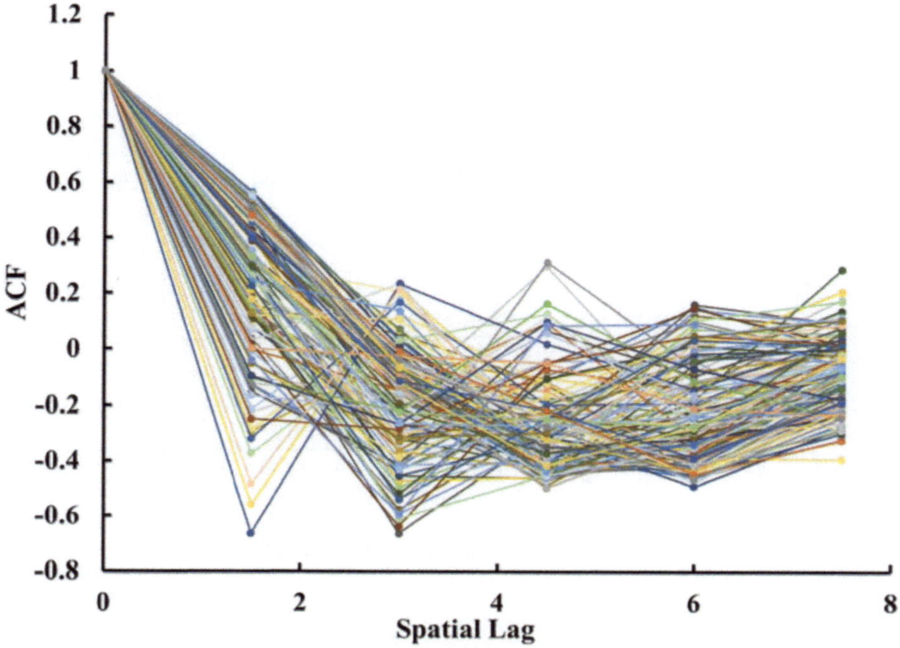

Fig. 10. Sample auto correlation function

By minimizing the mean square error between the estimated auto-correlation function and the ACM shown in Table 1, the parameter, *a*, of correlation structure has to be estimated from the standardized error. The fitted ACMs are shown in Fig. 11. The estimated correlation lengths for the exponential, squared exponential and second-order Markov ACMs are 1.65, 1.41 and 2.02 m, respectively. Squared exponential autocorrelation function is found to be best fitted ACM among the three theoretical ACMs. Hence the standardized error is recommended to be modelled as squared exponential random field for generation of one dimensional random field for errors. Therefore, the normalized residuals can be simulated as squared exponential random field. Knowing the correlation function, the SPT-*N* data can be simulated as follows:

$$N(z) = N_{avg}(z) + \sigma(z)\bar{w}(z) \tag{9}$$

Table 1. 1-D correlation functions

ACM	Correlation function	a	Correlation length(m)	RMSE	R^2
Exponential	$y = e^{-ax}$	1.212	2/a = 1.65	0.1898	0.8475
Squared exponential	$y = e^{-(x/a)^2}$	1.257	$a\sqrt{\pi}$ = 1.41	0.1796	0.8634
Second order Markov	$y = e^{-ax}(1 + ax)$	1.976	4/a = 2.02	0.1841	0.8565

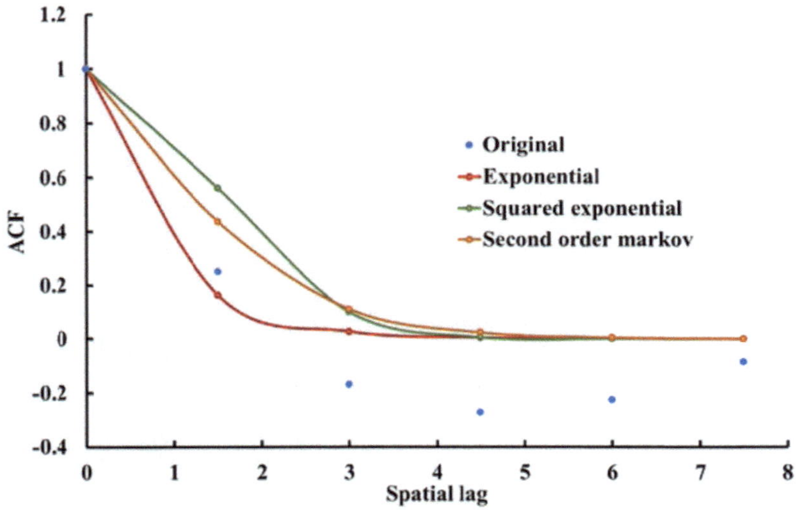

Fig. 11. Fitted auto correlation model

3 Conclusion

The primary objective of this paper is to report about the development of a stochastic model for characterizing the inherent spatial variation of SPT-*N* values obtained from field tests. The proposed stochastic model is developed from SPT borehole data collected from IIT Guwahati campus. Total 146 SPT-*N* borehole profiles collected from different construction companies associated with the construction of the campus are used for establishing this stochastic model. The *N*-value from SPT is considered as the summation of a deterministic part and a fluctuating component. The mean trend (deterministic part) characterizing the non-stationarity in data, is obtained from linear regression analysis. Residual part in the SPT-*N* data is modelled as a random field. The squared exponential correlation structure is found to be best fitted for this case. The scale of fluctuation (correlation distance) of the standardised error field is estimated from the fitted squared exponential correlation structure. The errors can be simulated easily from the estimated correlation structure and added to the deterministic part to simulate desire number of *N* profiles for further geotechnical analyses of the site.

References

Akkaya, A., VanMarcke, E.H.: Estimation of spatial correlation of soil parameters based on data from the Texas A&M University NGES. In: Probabilistic Site Characterization at the National Geotechnical Experimental Sites, ASCE, Reston, VA, pp. 29–40 (2003)

Stuedlein, A.W., Kramer, S.L., Arduino, P.H., Robert, D.: Geotechnical and random field modeling of desiccated clay. J. Geotech. Geoenvironmental Eng. ASCE **138**(11), 1301–1313 (2012)

Cafaro, F., Cherubini, C.: Large sample spacing in evaluation of vertical strength variability of clayey soil. J. Geotech. Geoenvironmental Eng. **128**(7), 558–568 (2002)

Campanella, R.G., Wickremesinghe, D.S., Robertson, P.K.: Statistical treatment of cone penetrometer test data. In: Proceedings of the 5th International Conference on Applications of Statistics and Probability in Soil and Structural Engineering, Vancouver, vol. 2, pp. 1011–1019 (1987)

DeGroot, D.J., Baecher, G.B.: Estimating autocovariance of in situ soil properties. J. Geotech. Geoenvironmental Eng. **119**(1), 147–166 (1993)

Elkateb, T., Chalaturnyk, R., Robertson, P.K.: An overview of soil heterogeneity: quantification and implications on geotechnical field problems. Can. Geotech. J. **40**, 1–15 (2003)

Fenton, G.: Random field modeling of CPT data. J. Geotech. Geoenvironmental Eng. **125**(6), 486–498

IS 2131: Method for standard penetration test for soils. Indian Standard Code of Practice (2002)

Jaksa, M.B.: The influence of spatial variability on the geotechnical design properties of a stiff, overconsolidated clay. Ph.D. thesis, Faculty of Engineering, University of Adelaide, Adelaide, Australia (1995)

Kulhawy, F.H., Roth, M.S., Grigoriu, M.D.: Some statistical evaluation of geotechnical properties. In: Proceedings of the 6th International Conference on Applications of Statistics and Probability in Civil Engineering, Mexico City, vol. 2, pp. 705–712 (1991)

Phoon, K.K., Kulhawy, F.H.: Characterization of geotechnical variability. Can. Geotech. J. **36** (4), 612–624 (1999)

Phoon, K.K., Quek, S.T., An, P.: Identification of statistically homogeneous soil layers using modified Bartlett statistics. J. Geotech. Geoenvironmental Eng. **129**(7), 649–659 (2003)

Popescu, R., Prevost, J.H., Deodatis, G.: Characteristic percentile of soil strength for dynamic analysis. In: Proceedings of the 3rd Conference on Geotechnical Earthquake Engineering and Soil Dynamics, Seattle vol. 2, pp. 1461–1471 (1998)

RaghuKanth, S.T.G., Dash, S.: Stochastic modeling of SPT N-value and evaluation of probability of liquefaction at Guwahati City. J. Earthq. Tsunami **2**(3) 175–196 (2008)

Uzielli, M.: Variability of stress-normalized CPT parameters and application to seismic liquefaction initiation analysis. Ph.D. dissertation, University of Florence (2004)

Uzielli, M., Vannucchi, G., Phoon, K.-K.: Random field characterisation of stress-normalised cone penetration testing parameters. Geotechnique **55**(1), 3–20 (2005)

Vanmarcke, E.H.: Random fields: Analysis and synthesis. MIT Press, Cambridge, MA (1983)

Scaling Factor for Generating P-Y Curves for Liquefied Soil from Its Stress-Strain Behavior

Suresh R Dash[1(✉)], Subhamoy Bhattacharya[2], and Praveen Huded[1]

[1] School of Infrastructure, Indian Institute of Technology Bhubaneswar, Bhubaneswar, India
srdash@iitbbs.ac.in
[2] Geomechanics, University of Surrey, Surrey, UK

Abstract. The technique of constructing a *p-y* curve relies upon the similarity between load-deformation characteristics of pile with the stress-strain behavior of the interacting soil as observed in many recent experimental studies. From a dimensional analogy of pile-soil interaction with flow type strain field, the strain distribution in soil due to pile deflection (*y*) can be considered proportional to the *y/D* ratio, where *D* is the diameter of the pile. When the depth of the construction is deep, the pile-soil interaction can be considered as a plain strain model. To obtain the *p-y* curve from stress-strain curve, scaling factors such as M_s and N_s are normally used. Although these scaling factors are well calibrated and are being used in practice since long for standard soil, its applicability and values for liquefied soil is very limited. This paper provides a simplified methodology of developing these scaling factors suitable for liquefiable soils by using a numerical procedure with virtual work principle.

1 Introduction

While analyzing the lateral response of pile foundation lateral pile-soil interaction is normally modelled as disjointed soil springs spaced at close intervals throughout the pile length, known as *p-y* springs. The backbone curve used in the *p-y* springs are called *p-y* curves. The *p-y* curve used in current practices have higher initial stiffness for both liquefied and non-liquefied soil. However for the liquefied soil the initial stiffness is very small and it increases with increasing displacement, as observed in many experimental and full scale studies (Rollins et al. 2005; Takahashi A. 2002; Dash 2010; Lamardi et al. 2017, etc.).

Following dimensional analogy of pile-soil interaction used by Osman and Boltan (2005), Klar (2008) and Klar and Osman (2008) for evaluating load deformation solutions, the strain field in the soil is represented by a single value over defined deformation zone. The average engineering strain (γ_{ave}) acts as a link between the strain of $\tau - \gamma$ (shear stress-strain) curve and deflection of the *p-y* curve (Bouzid et al. 2013). The scaling factor M_s equates the average shear strain with the *y/D* ratio. The γ_{ave} for the field case can be correlated with the shear strain of the element test of the soil sample. Once the displacement field is related to strain field, *p* can be calculated

H. Shehata and C. S. Desai (Eds.): GeoMEast 2018, SUCI, pp. 156–168, 2019.
https://doi.org/10.1007/978-3-030-01926-6_12

from stress in $\tau - \gamma$ curve of soil corresponding to γ_{ave} with a scaling factor, N_s (where, $p = N_s \tau_{\max} D$). The transformation of stress strain to *p-y* with the scaling factor M_s and N_s is shown schematically in Fig. 1.

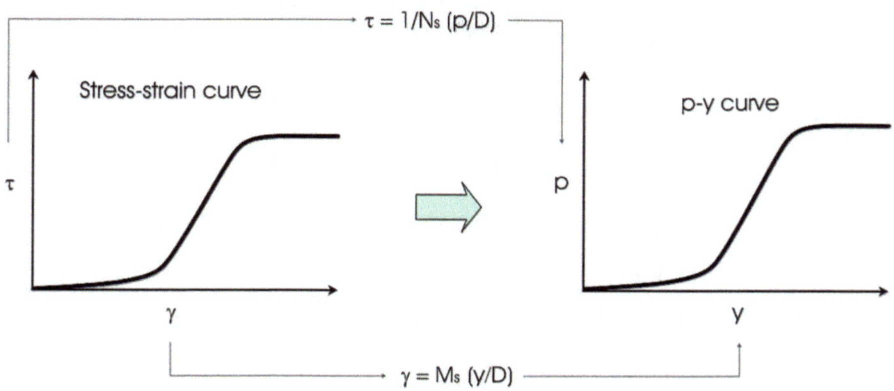

Fig. 1. Schematic showing the scaling of the p-y curve from the stress strain (Dash et al. 2017)

The scaling factor N_s is similar to the lateral bearing capacity factor (N_c) used to find the ultimate lateral bearing capacity of the pile. Similarly, M_s is similar to M_c which is used to relate a particular level of axial strain (ε_a) at half the maximum shear stress from the element test to define characteristic pile deflection (y_c) as $M_c \varepsilon_{50} D$, and is used while constructing *p-y* curve in clay. As per available literature, various values of N_c and M_c are given in Tables 1 and 2. The main aim behind the formulation of N_c and M_c was to obtain single value of p_u and y_c which may not be suitable for scaling of a full stress strain curve to get *p-y* curve. Also these factors cannot be used for the liquefied soil because of the limitation of the type of experiment and the materials used. Another drawback is that suggested values of N_c and M_c are not interrelated, but in reality strain field is related to the nonlinear stress-strain response. Hence N_s and M_s are seen only as scaling parameter, hence are different than N_c and M_c. Here, the effort has been made to establish the scaling factor N_s and M_s suitable for liquefied soil from mobilized strength principle, through virtual work approach.

2 Methodology

2.1 Virtual Work Approach for Choosing Scaling Factors, N_s and M_s

A virtual work approach was used to obtain the scaling factors, N_s and M_s. The pile-soil interaction at a particular depth was represented by a plane strain 2D model as shown in Fig. 2. Assuming the soil-pile system is initially in equilibrium, any incremental work done by pile will be equal to the incremental strain energy in the system. The following conditions were required to be satisfied in the system. (1) Pile deflection has to be

Table 1. Suggested values of bearing capacity factor N_c Suggested relationship between pile deflection (*y*) and soil strain

Reference(s)	N_c	Remarks
Matlock (1970)	$3+\frac{\sigma_z}{S_u}+J\frac{Z}{D}$ 3 at shallow depth 9 at deeper depth	Experimentally calibrated for different depth (z) J = 0.25 for Lake Austin clay, 0.5 for Sabine clay and Offshore clays in Gulf of Mexico
Reese et al. (1975)	$3+\frac{\sigma'_v}{(S_u)_{ave}}+2.83\frac{Z}{D}$	Experimentally calibrated
Reese and Welch (1975)	$3+\frac{(\sigma'_v)_{ave}}{(S_u)_{ave}}+0.5\frac{Z}{D}$	Experimentally calibrated
Gazioglu and O'Neill (1984)	$3+6\left(\frac{Z}{3D\left(\frac{E_pI_p}{E_sD_4}\right)}\right)$	Relative soil pile stiffness considered
Randolph and Houlsby (1984)	9.14 (smooth pile) 11.94 (perfectly rough pile) 10.5 (Suggested for practical use)	An exact solution based on plasticity theory for clayey soil
Kulhawy and Chen (1995)	9 (Driven pile) 12 (Drilled pile)	Suggested practical values
Einav and Randolph (2005)	18.566	A potential flow solution around a pile
Klar (2008)	14.52 (Rough pile) 10.06 (Smooth pile)	Analytical study based on mobilized strength design approach with elastic strain field around pile

Note: σ_z = Total overburden at depth z, S_u = Undrained shear strength at depth z, $(\sigma'_v)_{ave}$ and $(S_u)_{ave}$ = Average value over depth z

compatible with the strain distribution in the soil. (2) The stresses in the soil have to be in equilibrium with the total reaction force on pile.

Considering Tresca yield criterion, the incremental work done due to the displacement of inner disk in terms of maximum shear stress and shear strain is given by Eq. 1.

$$\delta W = \int_{vol} (|\tau_{\max}||\delta\gamma_{\max}|)dvol \tag{1}$$

Whereas, $\tau_{\max} = |\sigma_1 - \sigma_3|/2$ and $\delta\gamma_{\max} = |\delta\varepsilon_1 - \delta\varepsilon_3|$. Incremental work done was computed for a unit slice thickness and hence the unit of force. The equation for external and internal work done in the system is given by Eq. 2.

Table 2. Suggested relationship between pile deflection (y) and soil strain

Reference(s)	Axial soil strain (ε_a) to pile deflection, $y_c = M_{ce}\varepsilon D$	Remarks
McClelland and Focht Jr (1958)	$y = 0.5\varepsilon_a D$	Experimentally calibrated
Matlock (1970)	$y_{50} = 2.5\varepsilon_{50} D$ (Soft Clay)	Uses the relationship derived by Skempton (1951) for strip footing
Reese and Welch (1975)	$y_{50} = \varepsilon_{50} D$ (Stiff clay without free water)	Skempton (1951) concept modified with field test values
Reese et al. (1975)	$y_{50} = 2.5\varepsilon_{50} D$ (Stiff clay with free water)	Uses Skempton (1951) relationship and verified with field test data
Stevens and Audibert (1979)	$y_c = 8.9\varepsilon_{50}\sqrt{D}$	Calibrated with field test data with varying diameter
Gazioglu and O'Neill (1984)	$y_c = 0.8\varepsilon_{50}\sqrt{D}\left(\frac{E_p I_p}{E_s}\right)^{0.125}$	Considering the relative soil pile stiffness
Dunnavant and O'Neill (1989)	$y_c = 0.00638\varepsilon_{50} D\left(\frac{E_p I_p}{E_s L^4}\right)^{-0.875}$	Considering the relative soil pile stiffness
Klar (2008)	$y = \frac{1}{2.6}\gamma D$	Analytical study on fully plastic clay with an elastic strain field around pile

Note: y_c = Characteristic deflection, y_{50} = Deflection at half ultimate resistance, ε_a = Axial strain, ε_{50} = Axial strain at half maximum shear stress. γ = Engineering shear strain ($1.5\varepsilon_a$ in a triaxial tests)

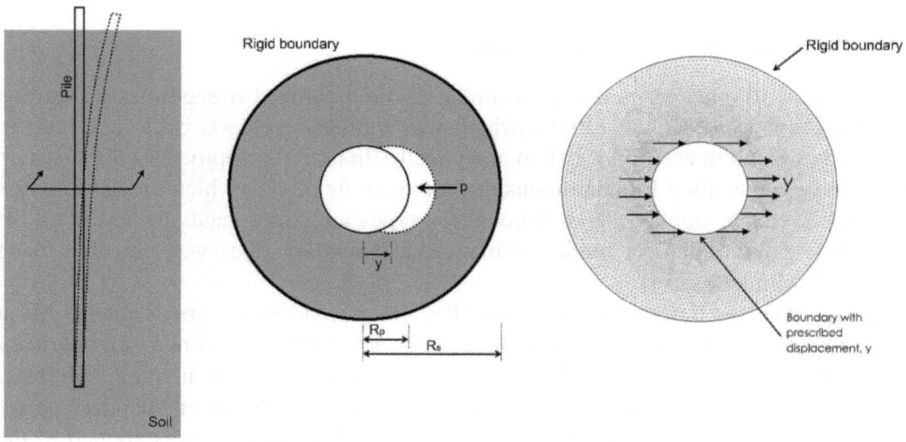

Fig. 2. Consideration of soil-pile interaction at a particular depth in FE analysis

$$p\delta y = \delta W \tag{2}$$

Where p, is the total reaction force over the boundary of the inner disk due to the incremental inner disk displacement of δy. Combining equation for N_s and Eq. 2, N_s can be written as Eq. 3.

$$N_s = \frac{\delta W}{\delta y \tau_{ave} D} \tag{3}$$

In Eq. 3, τ_{ave}, represents the average mobilized maximum shear stress, that is representative of the stress field around pile. The value of N_s depend on the extent of the fixed boundary chosen in the analysis. Displacements and stresses around a rigid disc have already been established by Baguelin et al. (1977). Displacement in both analytical and numerical analysis are model dependent. It was clear from the analysis, that the displacement to mobilize same lateral resistance increases with increasing soil boundary. For soil boundary tending to infinity the mobilized displacement will also tend to infinity. Consequently there will be infinite values of N_s and M_s as they rely upon the mobilized displacement and lateral resistance developed in soil. To obtain the rational values for N_s and M_s, it was required to find a reasonable boundary distance which allows to determine unique values for these coefficients. Using Upper bound theorem, τ_{ave} is sought in the considered soil domain that minimizes the N_s value for a known displacement of y. The boundary at which the lowest value of N_s occur can be considered as the effective deformation zone and the average shear strain mobilized (γ_{ave}) in the deforming soil due to pile deflection (y) can be obtained from the FE analysis as given by $\gamma_{\text{ave}} = \frac{\int_{vol} \gamma dvol}{\int_{vol} dvol}$ act as link between the strain of the $\tau - \gamma$ curve and the deflection of the p-y curve, and a scaling factor, M_s equates γ_{ave} with y/D as given by $\gamma_{\text{ave}} = M_s \frac{y}{D}$.

2.2 Development of Numerical Model

The virtual work method used here is similar to the mobilized strength design method used by Osman and Bolton (2005), which was further developed by Klar (2008) to represent pile-soil interaction problem analytically. In both the approaches the first step is to choose a boundary for continuous displacement field, over which the deformation mechanism can be applied, several boundary types and sizes needs to be tested. To study this, a finite element model with variable boundary sizes was prepared using COMSOL Multiphysics®.

A 2D pile soil interaction was modelled in a plane strain environment. Mesh optimization and some initial parametric studies were performed. A non-slip boundary was considered for the pile-soil interface, which correspond to a rough interface. Thereby the need of pipe material model was avoided and the inner boundary of the soil was considered as the pile surface. The pile displacement (y) was applied to the inner boundary of the soil. The exterior boundary was fixed and triangular mesh (Fig. 2) was adopted with Lagrange-quadratic elements.

2.3 Verification of Numerical Model

The numerical model developed has been tested for average strain field has been verified using a perfectly plastic material over an elastic field, with that of the numerical solution provided by Klar (2008). The solution has provided the extent of the fixed boundary = $2.85R_p$, which matches exactly with the solution of Klar (2008). The model is hence further used for liquefied soil.

2.4 Material Model for Liquefied Soil

Since the liquefied soil behaves like an incompressible fluid, an upper bound elastic solution was sought, where N_s was evaluated for different soil boundaries and the boundary at which N_s value was lowest considered as effective zone of influence. The simplified monotonic stress-strain curves of liquefied soil which shows hyper elastic behavior was used by having a strain dependent young's modulus material behavior. Six typical simplified stress-strain curve for liquefiable soils are adopted and their $E_s - \gamma$ relationship has been derived as shown in Fig. 3 a.

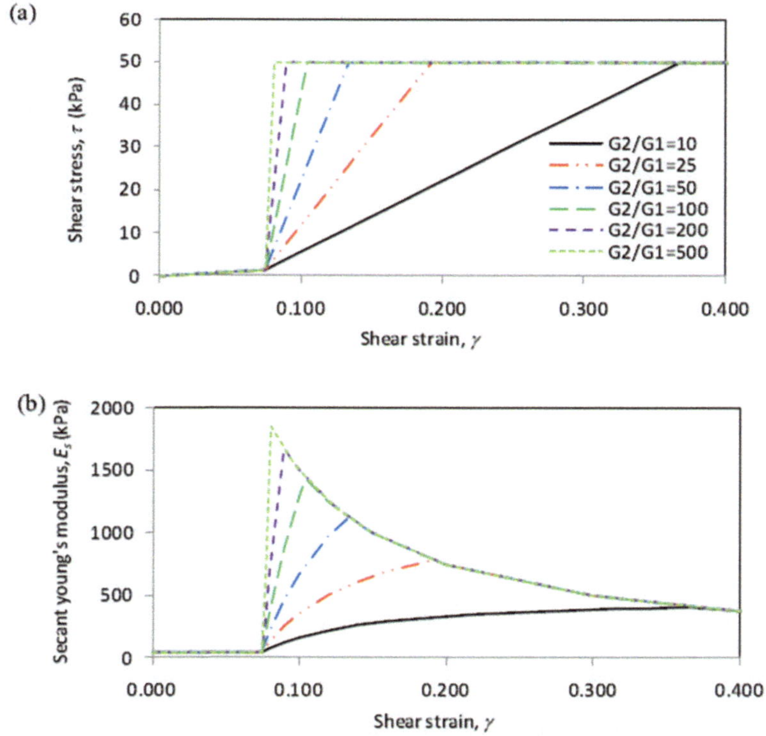

Fig. 3. Simplified model of liquefied soil used as input to the FE model in COMSOL

For all models, material nonlinearity of the soil was defined through a back bone curve of Young's modulus (E_s) versus engineering shear strain (γ). This backbone curve was obtained from the $\tau - \gamma$ curve of liquefiable soils adopted. In each step of the analysis, the maximum engineering shear strain was estimated and material's secant modulus (E_s) was modified accordingly as per the pre-defined $E_s - \gamma$ curve. An equivalent linear elastic model was used in COMSOL to include material nonlinearity.

3 Results and Discussion

Figure 4 shows the *p-y* curve and corresponding N_s and M_s values for the liquefiable soil model of $G_2/G_1 = 100$ for various boundary extent of the soil. When the stress condition in most of the soil elements was approaching the τ_{max} value (Fig. 3), the analysis was getting unstable. In practice, τ_{max} is a hypothetical value assigned to limit the stress-strain behavior of the soil. Hence, the convergence at this level doesn't significantly affect the results, as the zone of interest is the transition zone between G_1 to G_2. In the analysis, when the external fixed boundary was extended, it took more pile displacement to mobilize the second shear modulus G_2 of the soil. The *p-y* curves

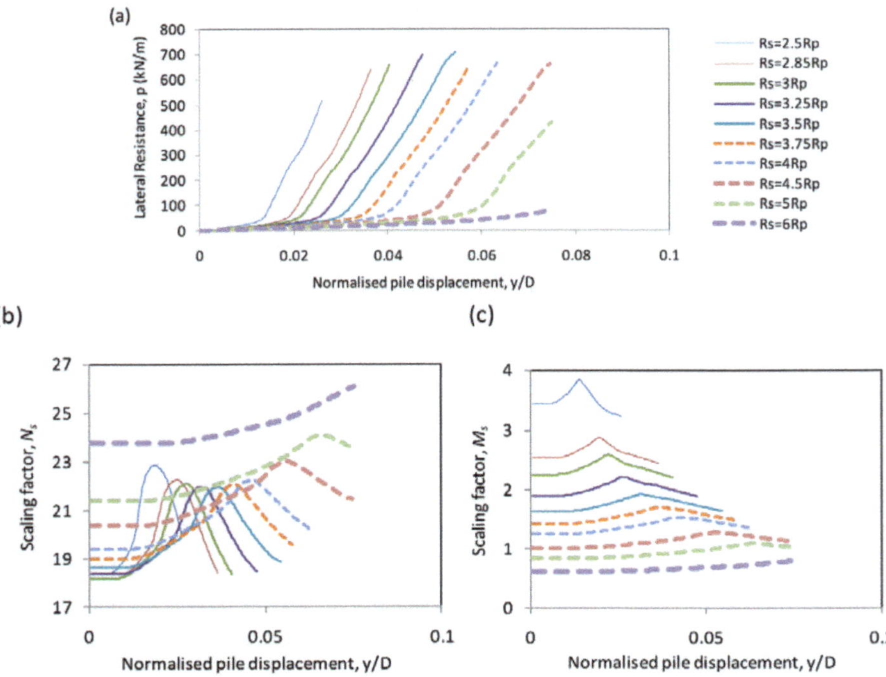

Fig. 4. (a) *p-y* curves for the liquefiable soil model of $G_2/G_1 = 100$ as obtained from numerical analysis by COMSOL for different soil boundary distances (R_s), and corresponding (b) N_s values (c) and M_s values

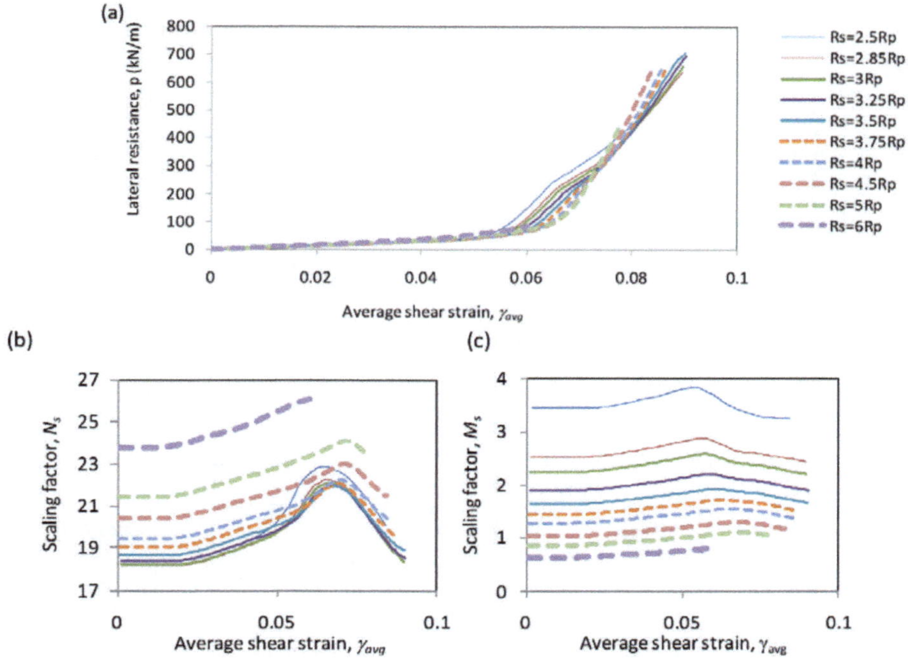

Fig. 5. (a) *p-y* curves for the liquefiable soil model of $G_2/G_1 = 100$ as obtained from numerical analysis for different soil boundary distances (R_s) plotted against the average shear strain in the soil, and corresponding (b) N_s (b) M_s values

derived in this way were highly dependent on the extent of the chosen boundary, and this affected the scaling factors as well. Following the analogy of average shear strain representing the deformation field in soil, the *p-y* curve and N_s and M_s curves with respect to average shear strain are determined as shown in Fig. 5.

From Fig. 5 it is clear that lateral soil resistance for all the boundaries with respect to average shear strain values roughly stack together. This shows that the average shear strain could probably be adequately used to represent the deformed state of the soil. N_s and M_s values were then studied with respect to average shear strain and plotted for different boundaries for all six liquefiable soil models in Fig. 6.

Considering the upper bound theorem, the boundaries corresponding to minimum N_s values were considered as the effective zone of influence and the corresponding values of N_s and M_s were recorded. It can be seen from the Figs. 6 and 7 that the effective zone is nonlinear and it increases with increase in nonlinearity in the system. When γ_{ave} was about 0.019, the nonlinearity in the system was noticeable. Although the average shear strain was still in initial elastic phase, some of the soil elements have already yielded and strained beyond 0.075 shear strain. In the very initial phase of straining, the soil boundary of $2.85R_p$ was acting as the effective zone, where most of

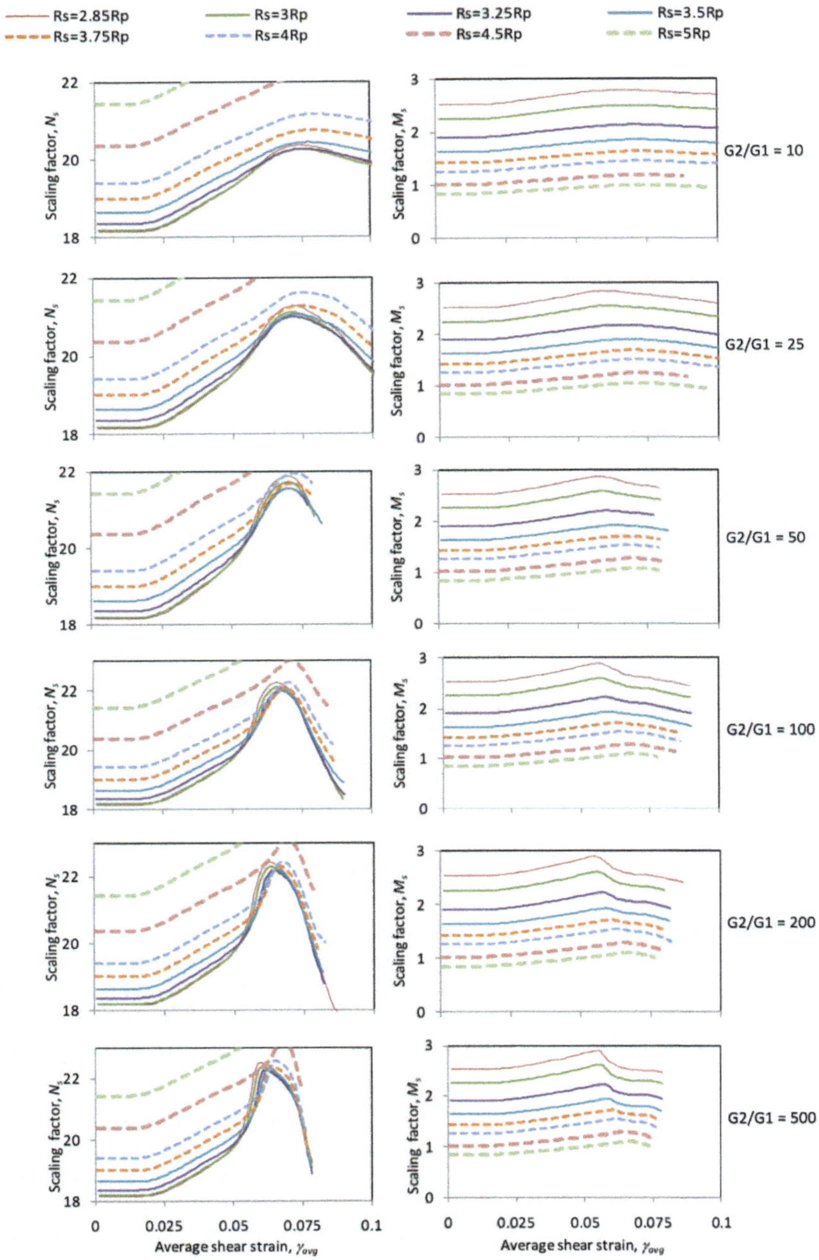

Fig. 6. Scaling factors N_s and M_s for liquefied soil with G_2/G_1 = 10, 25, 50, 100, 200 and 500

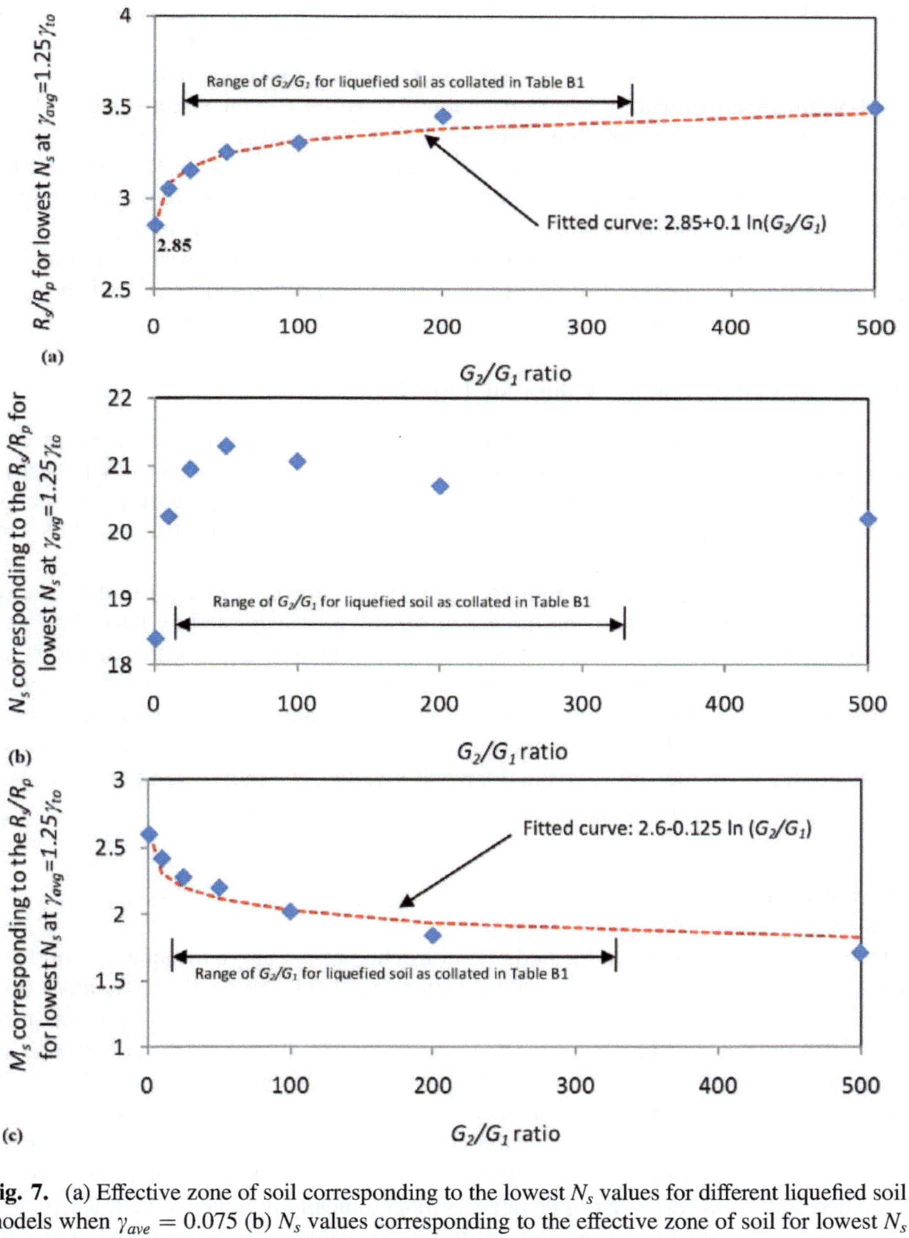

Fig. 7. (a) Effective zone of soil corresponding to the lowest N_s values for different liquefied soil models when $\gamma_{ave} = 0.075$ (b) N_s values corresponding to the effective zone of soil for lowest N_s (c) M_s values corresponding to the effective zone of soil for lowest M_s

the soil elements were linear elastic. When γ_{ave} exceeded 0.019 shear strain, the boundary of the effective zone of the soil corresponding to the minimum N_s shifted to a higher value. When γ_{ave} reached 0.075, it can be considered that the most of the soil elements have already yielded. Figure 7 show the variation of the effective soil zone (as a ratio of R_s/R_p) for different G_2/G_1 ratios considered in the model. The Fig. 7 show the logarithmic increment of effective zone with respect to the G_2/G_1 ratio.

For liquefied soil with higher G_2/G_1 ratio (i.e., stiffer G_2) the effective zone of soil was becoming larger higher. Similarly for $G_2/G_1 = 50$ and $G_2/G_1 = 200$, the effective zone of soil was $3.25R_p$ and $3.45R_p$ respectively. Figure 7 shows the M_s values for the effective zone of soil. The values of N_s for the effective soil zones were also changing for different G_2/G_1 ratios and different stress state as can be seen from Fig. 6. N_s was 18.16 for liquefied soil model for $G_2/G_1 = 100$, during initial elastic straining till $\gamma_{ave} = 0.019$, and it increased to 22 at $\gamma_{ave} = 0.066$, but decreased to 21.2 when $\gamma_{ave} = 0.075$ and then continued to decrease with further increase in strain.

3.1 Proposed N_s and M_s Values for Liquefied Soils

For design considerations, the lowest values of N_s and M_s can be considered as conservative as they give lesser strength and lesser stiffness to the *p-y* curve. The bearing capacity factor suggested by Martin and Randolph (2006) can be used as the scaling factor N_s for design purpose, which is 9.2 for smooth pile-soil interface and 11.94 for rough pile-soil interface.

The values of M_s obtained from FE analysis were lower compared to the solution by Klar (2008). For liquefied soil the use of $M_s = 2.6$ as suggested by Klar (2008) for plastic soil could be non-conservative as it will estimate larger stiffness. From the present study M_s was found to be nonlinear from the FE analysis, however, for practical applications and to keep the transformation of stress-strain to *p-y* curve linear, a single lowest value of $M_s = 1.87$, was proposed for fully liquefied soil. Although, the value of M_s increases when the soil is strained beyond yielding, the proposed value of M_s was kept 1.87 for the entire range of γ_{ave} to be in conservative side.

4 Conclusions

The present study provides a method of construction of *p-y* curve of liquefied soil from its stress-strain curve using two scaling factors N_s (load scaling factor) and M_s (displacement scaling factor). Numerical procedure with virtual work principle was used for evaluating these scaling factors. The numerical model developed is verified with the solution given in the literature and further used for liquefied soil model. Using FE analysis the approximate values of these scaling factors are derived for various shear modulus ratios of liquefied soil. Using these scaling parameters *p-y* curve can be proposed for the liquefiable soils subjected to monotonic loading, a condition when the liquefied soil flows past pile or pile deflects monotonically in liquefied soil. Keeping in view the conservativeness necessary in design the value of N_s is proposed as 9.2 for smooth pile-soil interface and 11.94 for rough pile-soil interface. Similarly, the value of M_s for fully liquefied soil is proposed as 1.87, for practical applications.

Acknowledgements. First author would like to acknowledge the financial help received from UK-India Education Research Initiative (UKIERI) for the DPhil study at the University of Oxford.

References

Baguelin et al.: The behavior of axially and laterally loaded single pile embedded in non-homogeneous soils. Geotechnique **21**(3), 309–326 (1977)

Bouzid et al.: Winkler Springs (p-y curves) for pile design from stress-strain of soils: FE assessment of scaling coefficients using the mobilized strength design concept. Int. J. Geomech. Eng. (Techno Press) **5**(5), 379–399 (2013)

COMSOL, A.B.: COMSOL Multiphysics user guide. Version-3.5a edition. COMSOL AB Stockholm, Sweden (2009)

Dash, S.R.: Lateral Pile Soil Interaction in Liquefiable Soils [Ph.D. Thesis]. University of Oxford, UK (2010)

Dash, S., et al.: A practical method for construction of p-y curves for liquefiable soils. Soil Dyn. Earthq. Eng. **97**, 478–481 (2017)

Dunnavant, T.W., O' Neill, M.W.: Experimental p-y model for submerged, stiff clay. Am. Soc. Civil Eng. J. Geotech. Eng. **115**(1), 95–114 (1989)

Einav, I., Randolph, M.F.: Combining upper bound and strain path methods for evaluating penetration resistance. Int. J. Numer. Meth. Eng. **63**(14), 1991–2016 (2005)

Gazioglu, S.M., O' Neill, M.W.: Evaluation of p-y relationship in cohesive soils. In: Analysis and Design of Pile Foundations: Proceedings of a Symposium Sponcered by the ASCE Geotechnical Engineering Division, San Francisco, California. October 1–5, 1984, pp. 192–213. American Society of Civil Engineers (1984)

Klar, A.: Upper bound for cylinder movement using elastic fields and its possible application to pile deformation analysis. Int. J. Geomech. **8**, 162 (2008)

Klar, A., Osman, A.S.: Load-displacement solutions for piles and shallow foundations based on deformation fields and energy conservation. Geotechnique **58**(7), 581–590 (2008)

Kulhawy, F.H., Chen, Y.: A thirty year perspective of Broms' lateral loading models, as applied to drilled shafts. In: Bengt Broms Symposium in Geotechnical Engineering (1995)

Martin, C.M., Randolph, M.F.: Upper bound analysis of lateral pile capacity in cohesive soil. Geotechnique **56**(2), 141–145 (2006)

Matlock, H.: Correlation for design of laterally loaded piles in soft clay. In: Offshore Technology Conference (1970)

McClelland, B., Focht Jr., J.A.: Soil modulus for laterally loaded piles. Trans. ASCE **123**, 1049–1063 (1958)

Osman, A.S., Boltan, M.D.: Simple plasticity-based prediction of the undrained settlement of shallow circular foundations on clay. Geotechnique **55**(6), 435–447 (2005)

Randolph, M.F., Houlsby, G.T.: Limiting pressure on a circular pile loaded laterally in cohesive soil. Geotechnique **34**(4), 613–623 (1984)

Reese, L.C., Welch, R.C.: Lateral loading of deep foundations in stiff clay. J. Geotech. Eng. Div. **101**(7), 633–649 (1975)

Reese, L., et al.: Field testing and analysis of laterally loaded piles in stiff clay. In: Offshore Technology Conference (1975)

Rollins, K.M., et al.: Lateral resistance of a full scale pile group in liquefied sand. J. Geotech. Geoenvironmental Eng. **131**, 115–125 (2005)
Stevens, J.B., Audibert, J.M.E.: Re-examination of p-y curve formulations. In: Proceedings 11th Annual Offshore Technology Conference, p. 397 (1979)
Takahashi, A.: Soil-pile interaction in liquefaction induced lateral spreading of soils. [Ph.D. Thesis], Tokyo Institute of Technology, Japan (2002)

Finite Element Analysis of Sub-surface Settlements and Pile-Tunnel Interaction

Animesh Sharma(✉)

Geotechnical Engineer, AFCONS Infrastructure LTD, Kolkata East West Metro UG1, Kolkata, India

Abstract. With trend of growing urbanization, use of Rapid Mass Transit Systems using underground tunnels has exponentially increased. Due to urban locale of network of underground transit system, often high rise buildings are found in close vicinity and with growing rate of underground urban infrastructure, it can be further envisaged that soon underground construction will impact existing underground infrastructures as well. Underground construction exhibits ground deformations in ground which impacts the existing infrastructure. Current study is based on assessment of sub-surface settlements and impact on pile foundations due to tunneling using Finite Element Analysis software i.e. PLAXIS 2D. The study extends to comparison with semi-empirical analyses using relevant semi-empirical methods in total stress conditions. Numerical Regression has been undertaken for compared trends for both sub-surface settlements and pile-tunnel interaction in terms of surface settlements in total stress condition. The resulting trends have been further compared with the existing case studies and relevant literatures and are found to be in fair agreement.

1 Introduction

Due to exponential increase in urbanization, there comes a parallel growth in urban infrastructure. Due to dimensional limitation of surface transportation system, exploration of underground space for systematic mass transit system has increased and because of the urban locale, it is of utmost importance to consider its impact on existing infrastructure. With further advances in infrastructure, soon tunneling under existing tunnel or near existing deep foundation element is envisaged to be a regular design aspect. With that under consideration, current study focuses on two important aspects of urban tunneling; assessment of sub-surface settlements and pile-tunnel interaction. The study undertakes finite element analysis using PLAXIS 2D for estimation of both aspects; sub-surface settlements and pile-tunnel interaction and compares it with various relevant methods developed so far for estimation of both aspects. The study further tries to establish and validate correlations for assessment of pile settlements using surface settlements in total stress conditions.

H. Shehata and C. S. Desai (Eds.): GeoMEast 2018, SUCI, pp. 169–181, 2019.
https://doi.org/10.1007/978-3-030-01926-6_13

2 Surface and Sub-surface Settlements Due to Bored Tunnels

2.1 Surface Settlements

Ground movements are inevitable aspect of tunneling. Urban Tunneling requires a detailed analysis for assessment of ground movements and consequent ways to minimize them. Tunneling is basically a form of excavation which causes relaxation of ground and hence changes in stresses during the process. The changes in stresses further causes ground a movement which then impacts the buildings in vicinity. Earliest studies for assessment of ground movements can be traced to Peck (1969) whereby empirical formula was suggested based on the field observations. Equation 1 and Figure 1 depicts the semi-empirical formula and Gaussian settlement trough suggested by Peck. The equation was over time further upgraded by various researchers based on the further field and centrifugal model tests. Further hypotheses were presented by Sagaseta (1987), Verrujit and Booker (1996), Addenbrooke et al. (1997) using various analytical and numerical approaches including effect of various parameters, different soil models and considering soil structure interaction.

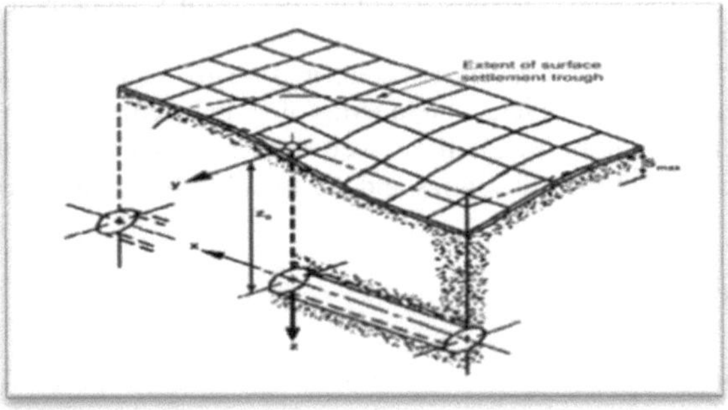

Fig. 1. 3D Representation of generated settlement trough due to tunnel (Yeates 1985)

$$s = Smax.\exp\left(-\frac{y^2}{2i^2}\right) \tag{1}$$

$$Vs = \sqrt{2\pi} * k * z * Smax \tag{2}$$

Where, s = settlement at any horizontal distance, y
S_{max} = maximum settlement; V_s = Volume of settlement trough
Z = depth to tunnel axis; y = horizontal distance
i = point of inflexion
k = trough width parameter (assumed as 0.5 for clayey soils and 0.3 for sandy soils)

2.2 Sub-surface Settlements

With growing dimensional limitations of surface transportation, underground exploration has exponentially risen. With this continued rise of urban underground exploration, soon it will be very important to consider the impact of tunneling on existing underground structures like tunnels as well. Hence, consideration of sub-surface settlements due to tunneling will be a necessary design impact.

Only few studies have been undertaken so far for the assessment of Sub Surface settlements due to tunneling. Earliest studies can be traced to Atkinson and Potts (1979), whereby empirical relations were suggested based on physical centrifugal test results. Further, Mair et al. (1993) suggested empirical formula based on field observations in London Clay. There were still various limitations to the empirical relations such as their applicability in varying ground conditions. Further Loganathan and Poulos (1998) suggested equation based on closed form solution which was basically a further modification to the technique of Verujit et al. (1996) and Sagaseta (1987).

Mair et al. in 1993 suggested the following theory based on field interpretations and physical centrifugal modeling in clayey soils. Figure 2 (a) depicts the transverse trend of sub-surface settlements observed using semi-empirical approaches and physical centrifugal tests as suggested by Mair et al. 1993, whereas Fig. 2 (b) shows comparative sub-surface settlements as compared with Loganathan approach and Mair approach.

$$kz = 0.175 + \frac{0.325\left(-\frac{z}{z0}\right)}{1 - z/z0} \tag{3}$$

$$Iz = Kz.(z - z0) \tag{4}$$

Equations 3 and 4 were the modifications suggested by Mair et al. (1993) for trough width parameter of Eq. 1 for assessment of sub-surface settlements where z0 is depth of sub-surface settlement in clayey soil conditions.

Further closed form approach was suggested by Loganathan et al. (1998) for estimation of sub-surface settlements. Equation 5 details the equation suggested by Loganthan et al. (1998) for estimation of sub-surface settlements.

$$\begin{aligned} Uz = \acute{\varepsilon}0R2\,\{(-(z-H)/(x2+(z-H)2+(3-4v)(z+H)/(x2+(Z+H) \\ 2-2z(x2-(z+H)2/(x2+(z+H)2)2\}.\exp\{-[1.38\times 2/(H\cot\beta+R2)+0.69z2/H2 \end{aligned} \tag{5}$$

Where. R = Radius of tunnel, z = Depth of tunnel, H = Depth of sub-surface settlement, V = poissons ratio, X = horizontal distance.

For the current study, the results of the semi-empirical methods as mentioned in Eqs. 3, 4 and 5 have been compared to the results of the Finite Element Analysis using PLAXIS 2D. The sub-surface settlements have been further determined as a function of surface settlements and are further correlated with pile settlements in context with geotechnical parameters and distance from tunnel axis.

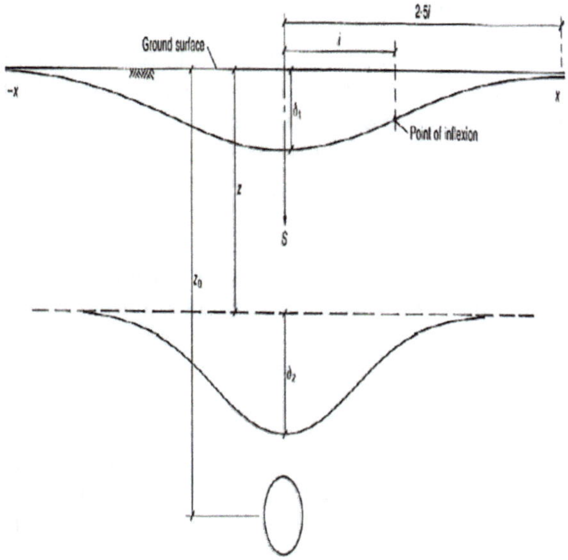

(a)

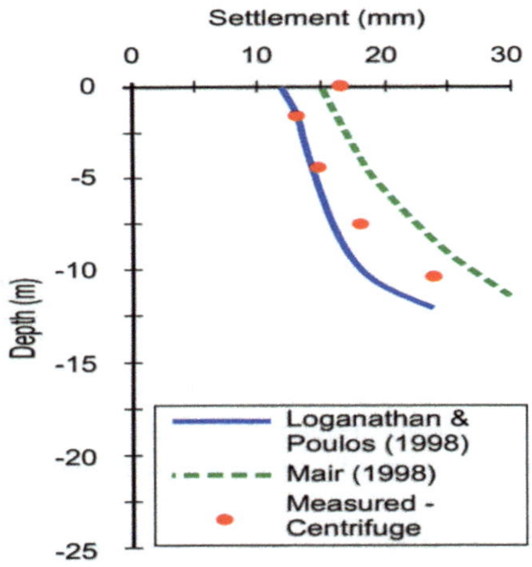

(b)

Fig. 2. (a) Trasnverse trend of sub-surface settlements (b) Comparison of trends of sub-surface settlements as obtained using various relevant methods

3 Pile-Tunnel Interaction

Often many High Rise Buildings and Heavy Structures like Bridges etc. are located in close vicinity to the tunnel. The ground deformations induced due to tunneling then becomes important; hence it is vital to study the impact of tunneling on the pile foundations. Earliest studies relating to the Impact of tunneling on Pile Foundations can be traced to Loganathan and Poulous (1999) impact of tunneling piles in clayey soil was assessed. Further Jacobsz et al. (2001) based on centrifuge model tests suggested pile settlements according to relative location of tunnel with respect to the pile foundation. Figure 3 (a) shows the considered zones based on location of tunnel with respect to pile foundation and assessed pile settlements with respect to surface settlements.

Further studies were undertaken by Poulos et al. (2006) for assessment of tunnel operation induced pile settlement in different soil conditions and their corresponding axial capacity loss. Figure 3 (b) shows the base case considered for the analysis and subsequent settlements were correlated to the surface settlements in different initial stress conditions using approximate closed form solutions.

Further Loganathan et al. suggested design charts by undertaking a detailed parametric study to examine the impact on piles. Based on the observed results, design charts along with parametric corrections were proposed for assessment of impact on piles. A base case was considered for analysis and further design charts were proposed in order to assess the impact on pile.

Another empirical approach considered for assessment of impact on piles in the current study is 2/3 depth approach. According to the approach, the settlements at the pile head are equivalent to the sub-surface settlements at 2/3 times the tunnel depth. However the source of the considerations of approach is unknown to the author, yet it the results have been considered and compared with the results of the Finite element analyses. Earlier in 2011, Mike Devriendt et al. considered 2/3 Approach for analysis and compared the pile head displacements to the results of CTRL case study and found to show a good agreement.

4 Finite Element and Regression Analysis for Sub Surface Settlements and Pile-Tunnel Interaction

Finite element analysis has been undertaken for assessment of surface, sub-surface and pile settlements using finite element software i.e. PLAXIS 2D. Since the geometry of tunnel was considered as uniform, axis symmetry model was considered for analysis. The mesh was considered as fine. The geometry considered was symmetrical in nature. Standard fixities were considered for the model with full fixity at bottom ($U_x = U_y = 0$) and roller condition at vertical sides ($U_x = 0$). Hardening Soil model was used for the simulation citing the stress dependent accounting of stiffness modulus in it. Tunnel lining was simulated using the tunnel element provided in PLAXIS 2D. For the analysis, 1% volume loss was assumed. Pile was modeled using plate element. Table 1 lists the default geotechnical, pile and tunnel properties assumed for the analysis, which

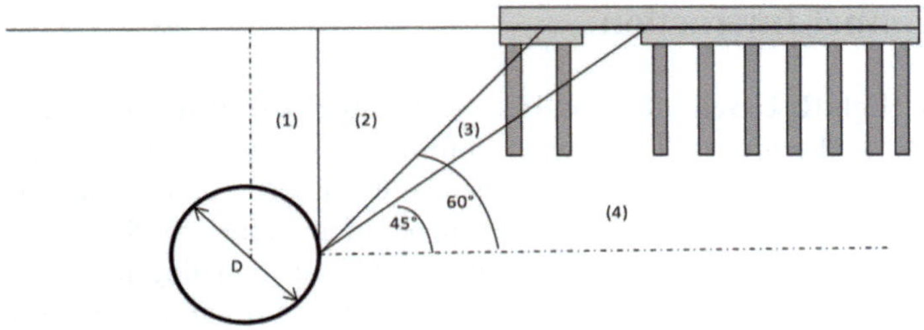

Expected Pile Settlement Compared to Ground Settlement (after Jacobs et al 2002)		
Pile Toe Zone	**V_L <= 1%**	**V_L > 1%**
(1)	Similar or Less	More
(2)	Similar or Less	More
(3)	Similar or Less	Approximately Equal
(4)	Less	Less

(a)

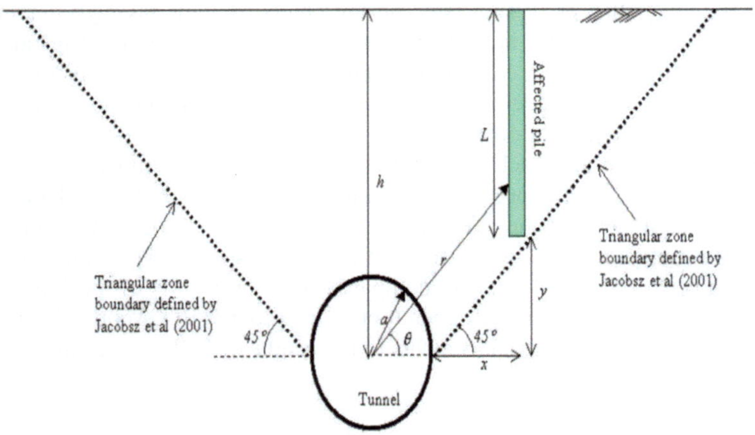

(b)

Fig. 3. (a) Zones of Pile-Settlement as defined by Jacobs et al. (a) Pile settlement base case as considered by Poulous et al

were varied to study the effect on ground deformations. The parameters considered were similar to that considered in base case by Loganathan et al. for detailed parametric study for design charts and considered by Poulos (2006) using approximate closed form solutions.

4.1 Sub-surface Settlements

Sub-Surface settlements assessed using semi-empirical approach and finite element analysis are presented in Fig. 4 shows and depicts the comparison of maximum sub-

Table 1. Geotechnical, pile and tunnel parameters

Parameter	Symbol		Value	Unit
Cohesion	c'	:	0.5	kPa
Undrained cohesion	Cu	:	75	kPa
Angle of internal friction	Φ'	:	30	Degrees
Hardening soil model	E_{50}^{ref}	:	12000	kPa
	E_{oed}^{ref}	:	8000	kPa
	E_{ur}^{ref}	:	25000	kPa
Shear modulus	G_{ref}	:	4000	kPa
Co-efficient of earth pressure	k0	:	0.5	
Bulk density	ϒ	:	19	kN/m^3
Dry density	ϒd	:	17	kN/m^3
Tunnel parameters				
Tunnel diameter	D	:	6	m
Axial stiffness	EA	:	9000000	kNm/m
Bending stiffness	EI	:	32000	kNm^2/m
Segment thickness	T	:	0.3	m
Cover to tunnel	C	:	12	m
Pile parameters				
Axial stiffness	EA	:	1E6	kNm/m
Bending stiffness	EI	:	1E5	kNm^2/m
Pile diameter	d	:	01	m

surface settlements along the tunnel centre line axis in total stress conditions. Figure compares the Settlement Ratio i.e. the ratio of sub-surface settlements/surface settlements to the ratio of sub-surface depth of settlement/tunnel axis depth. Evident from results depicted Fig. 4 that the trends from both methods show good degree of similitude with each other and thereby validates the numerical analysis.

Figure 5 shows the transverse trend of settlement ratio i.e. sub-surface settlements to surface settlements at various depths with respect to horizontal distance from tunnel axis as obtained according to Loganathan approach. It can hence be established from the observed trend that the ratio of sub-surface settlements to surface settlements are function of tunnel depth. With increase in depth, although the magnitude of sub-surface settlement increases, but consequently decreases the point of inflection or zone of influence of sub-surface settlement as compared to the surface settlements. In other words it can be said that the sub-surface settlements get more concentrated towards the tunnel centerline with respect to increase in depth. The sub-surface settlement increases to a ratio of 1.5–2.5 times the surface settlements along the tunnel axis and maximizing at tunnel crown level. Figure 6 compares the transverse trends from Loganathan et al. with the results of finite element analysis. Figure 6 (a) depicts a screenshot of typical transverse pattern of sub-surface settlements as obtained from Finite Element Analysis and Fig. 6 (b) represents transverse trend as obtained according to Loganathan approach thereby representing a similar trend.

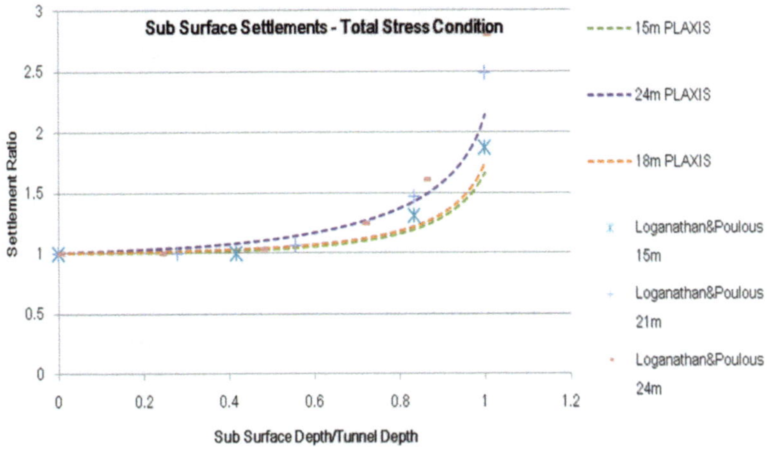

Fig. 4. Comparison of Sub-Surface settlements along the tunnel axis in total stress condition using PLAXIS and Loganathan Approach

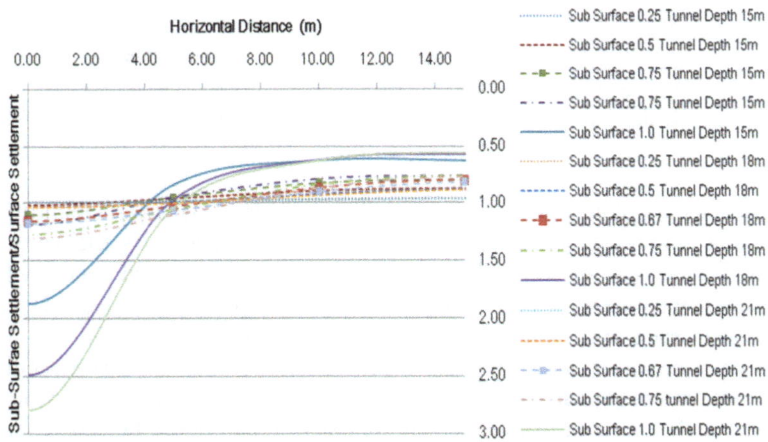

Fig. 5. Transverse graphs of sub-surface settlements (Loganathan approach)

The sub-surface settlements from both Finite Element analysis and Loganathan approach are compared and are depicted in context with their ratio to surface settlements with their horizontal distance from tunnel alignment in Fig. 7. The results from both approaches show good agreement and forms definite trend for interpolation and interpretation of sub-surface settlements in total stress/undrained conditions in terms of surface settlements.

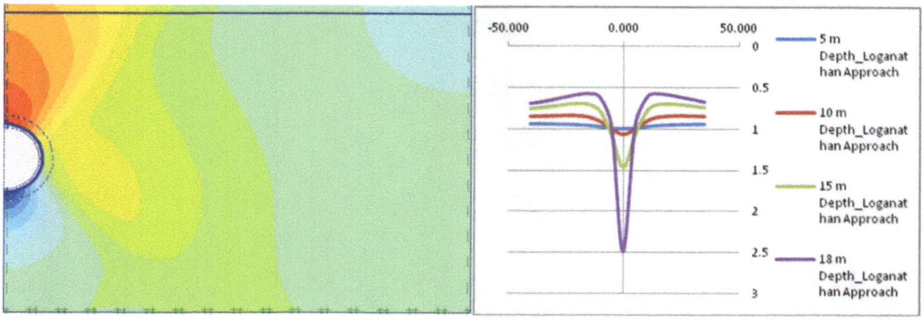

Fig. 6. Typical transverse trend of sub-surface settlements using (a) Finite Element Analysis (b) Loganathan et al. approach

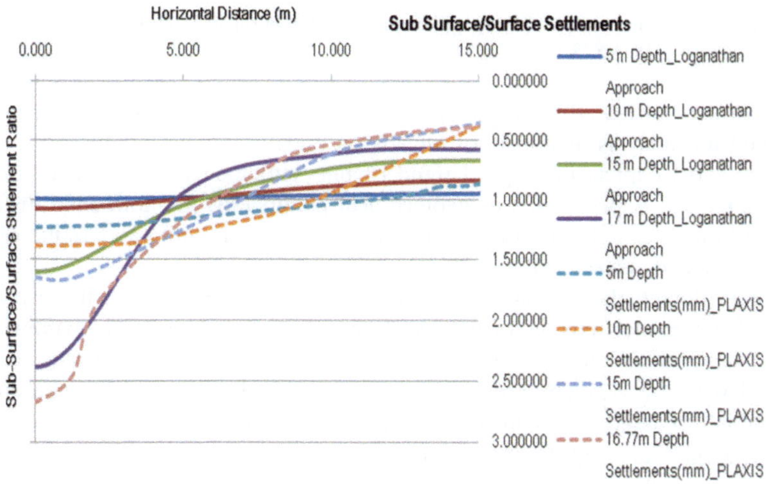

Fig. 7. Comparison of transverse sub-surface/surface settlement ratio

4.2 Pile Settlements

Figure 8 (a) shows the model used for finite element analysis for assessment of pile settlements due to bored tunnel. Table 1 details the geotechnical, tunnel and pile parameters considered for the analysis. For current study, only short piles i.e. case where length of pile is less than the depth of tunnel axis is considered. The horizontal distance of pile from tunnel axis has been considered as a varying parameter and its effect has been studied. Since the comparisons have been made to the results from Loganathan approach, total stress/undrained conditions have been considered.

Parameters such as tunnel diameter, tunnel depth, diameter of pile etc. considered for finite element analysis and 2/3 sub-surface settlements are considered in accordance to that of the base case of Loganathan approach. The finite element model used for

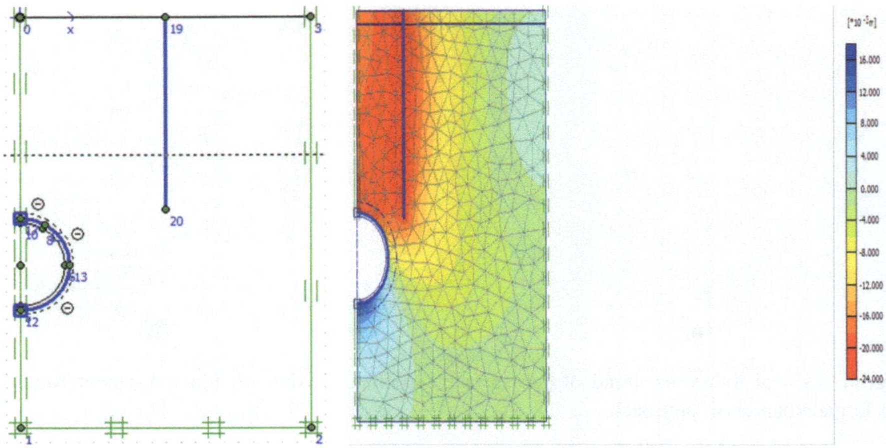

(a) (b)

Fig. 8. (a) Finite element model for pile-tunnel interaction (b) Typical trend of vertical displacement using PLAXIS

analysis of sub-surface settlements as mentioned in previous section has been further considered for analysis of impact of tunnels on piles.

Figure 8 (b) depicts the typical trend of pile settlements as obtained from finite element analysis. Aspects such as geotechnical parameters, the horizontal distance of pile from tunnel axis etc. are then varied in order to study and validate the observed trends of pile and surface settlements.

Figure 9 depicts the comparison of ratio of Pile Head settlement with respect to its distance from tunnel axis in context with variation in geotechnical parameters, as obtained from finite element analysis and Loganathan approach. The results are further compared with the 2/3 sub-surface settlements as obtained using the semi-empirical approach as discussed earlier in study in respect with horizontal distance from tunnel axis. Apparent from Fig. 9 that the trend of results obtained from finite element

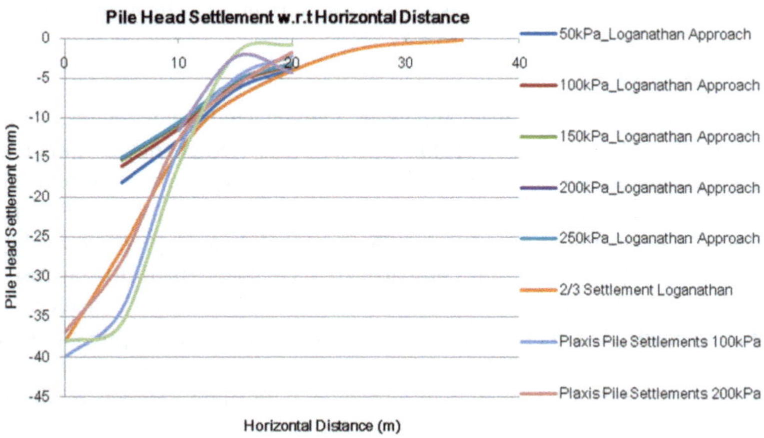

Fig. 9. Pile head settlement with respect to horizontal distance from tunnel axis

approach and Loganathan approach concur with the trend of 2/3 sub-surface settlements and provides a good range for assessment.

Based on the above observed trends, the application of sub-surface settlements can be validated for estimation of pile settlements. It is evident from the results of previous study that the pile settlement is a function of geotechnical parameters to a great extent.

The obtained results using finite element analysis are found to be in accordance with the results obtained using different approaches like physical centrifuge modeling which were generalized by Jacobz et al. in 2001 and further validated in context with geotechnical capacity loss using boundary element approach by Poulos in 2006. The results are also compared to the design charts suggested by Loganathan et al. for estimation of impact of tunneling on piles. The results are found to be validated with the analytical plane strain model developed by Bezuijen et al. (2014) which was validated using several case studies.

The trend of settlements obtained using various approaches are then further correlated with the surface settlements as obtained by Finite element analysis and theoretical methods. Figure 10 below details the trend of Pile/surface settlements as obtained for various tunnel axis depths using Finite Element Analysis, Loganathan Approach and 2/3 Sub-surface settlements. The results obtained are comparable with each other. Since depth of tunnel is considered a varying parameter, a generalized trend can be developed for estimation of pile settlements.

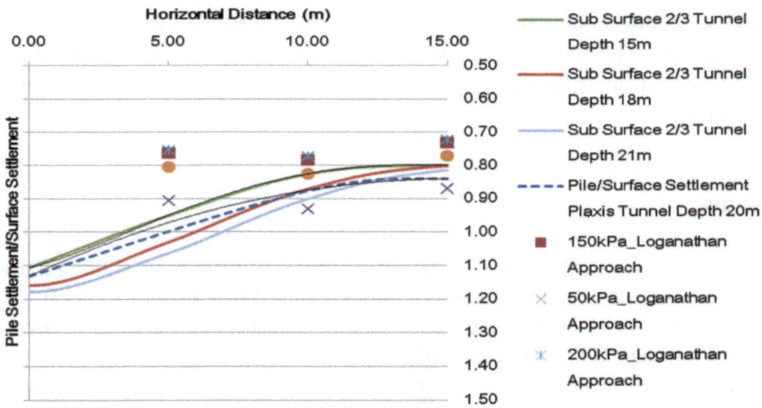

Fig. 10. Comparison of pile/surface settlements from Finite element, 2/3 sub-surface and Loganathan approach

Regression Analysis for Estimation of pile settlements based on Surface Settlements

$$p = 0.2681 * x - 1.5D \quad (6)$$

Where p = pile/surface settlement
x = horizontal distance of pile from tunnel axis, D = Tunnel Diameter

The above results obtained from finite element analysis validate the conventional methods of assessments of pile settlements and provides a generalized trend factor for assessment of pile settlement (short piles) in terms of surface settlements based on horizontal distance from tunnel in total/undrained stress conditions.

5 Conclusions

With growing rate of urban infrastructure, soon consideration of assessment of impact of tunneling on underground structures like existing tunnels will be an inevitable design aspect. Current study undertakes finite element analysis of sub-surface settlements using PLAXIS 2D in Total stress conditions and compares the results with relevant results and semi-empirical techniques like Mair et al. (1993) and Loganathan et al. Upon comparison, the trends of results are found to be in accordance with each other. The sub-surface settlements are assessed and represented as functions of surface settlement in transverse direction and with respect with tunnel depth. Consideration of this will allows us to interpret or interpolate the sub-soil movements during tunneling and hence enable us to envisage the impact on underground structure. Finite Element Analysis for Pile-Tunnel Interaction has been undertaken and has been consequently been compared with the results of closed form solutions and boundary element theory. Another approach, i.e. 2/3rd Sub-Surface Settlements approach i.e. sub-surface settlements of soil at 2/3 times the tunnel depth from ground level as been considered for the comparison with pile head settlements obtained from finite element analysis and relevant approaches. The results concur with each other and essentially allow us to estimate the impact on short piles due to tunnels based on surface settlements in total stress condition.

References

Addenbrooke, T., Potts, D.M., Puzrin, A.M.: The influence of pre-failure soil stiffness on the numerical analysis of tunnel construction. Geotechnique **47**(3), 693–712 (1997)

Atkinson, J.H., Potts, D.M.: Subsidence above shallow tunnels in soft ground. J. Geotech. Eng. Am. Soc. Civil Eng. **GT4**, 307–325 (1979)

Poulos H.G.: Pile Settlement zones above and around tunneling operations. Aust. Geomech. **41** (1), 81–90 (2006)

Jacobsz, S.W., Standing, J.R., Mair, R.J., Soga, K.: Tunneling effects on driven piles. In: Proceedings of the Conference on Response of Buildings to Excavation-Induced Ground Movements, London, UK (2001)

Loganathan, N., Poulos, H.G., Xu, K.J: Ground and Pile group responses due to tunneling. Soil Found. **41**(1), 57–67 (2001)

Loganathan, N., Poulos, H.G.: Analytical predictions of tunneling induced ground movements. Geotech. Eng. J. Am. Soc. Civil Eng. **124**(9) (1998)

Mair, R., Taylor, R., Bracegirdle, A.: Sub-surface settlement profiles above tunnels in clay. Geotechnique **43**(2), 315–320 (1993)

Peck, R.B.: Advantages and limitations of the observational method in applied soil mechanics. Geotechnique (ICE, London) **19**(2), 171–187 (1969)

Sagaseta, C.: Analysis of undrained soil deformation due to ground loss. Geotechnique **37**, 301–320 (1987)

Dias, T.G.S., Bezuijen, A.: Pile-tunnel interaction: a conceptual analysis Deltares, Delft, The Netherlands Geotechnical Aspects of Underground Construction in Soft Ground. 2014 Korean Geotechnical Society, Seoul, Korea. ISBN 978-1-138-02700-8

Verruijt, A., Booker, J.R.: Surface settlements due to deformation of a tunneling an elastic half plane. Geotechnique **46**(4), 753–756 (1996)

Evaluation of Design Parameters of Near Embankment Underground Tunnel Structure by Numerical Analysis of a 2D Plain Strain Model

A. R. Rahman[1(✉)] and R. Ahsan[2]

[1] Institute of Water Modelling (IWM), Dhaka, Bangladesh
[2] Department of Civil Engineering, Bangladesh University of Engineering and Technology (BUET), Dhaka, Bangladesh

Abstract. Now-a-days, traffic congestion is a major problem in many cities. Due to excessive pressure on overland roadways, underground subway or roadway is regarded as an alternative solution to accommodate this huge traffic. For this reason, construction of a suitable underground tunnel is a good alternative. Because the urban tunnel construction is normally carried out in difficult and crowded city areas, one of major design and construction challenges lies in minimizing the induced ground movement, as it poses potential threats to nearby buildings and utilities along the tunnel alignment. Most of the previous works were done to study the impact of the tunnel excavation on adjacent buildings which is of major interest for tunneling construction in urban areas, due to the high interaction between tunneling and existing structures. In context of the previous studies, present study aims at investigating the design parameters of an underground tunnel which is constructed in such a zone where a residential area is being developed far away from the business center, for example, near any river embankment, where new land is being developed using sand filling taken from the dredged materials from the bed of the rivers. The response of the nearby structures (displacement, bending moment, shear force) will not only be dependent on the construction sequence of the tunnel, but also will be dependent on the change in water level near the embankment as these embankments are subjected to a certain water head difference due to difference in water levels on both sides of the embankment causing seepage of water through the embankment and generating uplift forces in adjacent structures. Such phenomena may also induce lateral displacements to nearby tunnels, buildings and if overlooked, can cause severe structural damage at their base. In this particular study, a simple 2D plain strain model was developed in PLAXIS-2D. Various scenarios were simulated by construction of tunnel and varying the water level on one side of the embankment. The behavior of both nearby tunnel and building was observed in terms of generated maximum displacement, moment and shear force. Simulated results showed both increase and decrease in the parameters mentioned above for varying thickness of tunnel, vertical distance of center of tunnel from existing ground level, distance between tunnel center and center of structure and varying water head difference on both sides of the embankment.

H. Shehata and C. S. Desai (Eds.): GeoMEast 2018, SUCI, pp. 182–198, 2019.
https://doi.org/10.1007/978-3-030-01926-6_14

1 Introduction

Bangladesh is a riverine country. It is criss-crossed by hundreds of rivers. To protect the country side from flood waters, embankments are constructed along many of these rivers. To match with the population growth, many residential areas are now being developed on sides of these river embankments. These factors in turn creating mass traffic problems and necessitating new road ways and alternate mode of transport facilities. However, due to lack of available space and excessive pressure on overland roadways, underground subway or roadway is regarded as an alternative solution to accommodate this huge traffic. For this reason, construction of a suitable underground tunnel is a good alternative Many works currently undertaken in all parts of the world in the field of tunnelling are related to the construction and extension of underground network in big overpopulated cities [1]. Because the urban tunnel construction is normally carried out in difficult and crowded city areas, one of major design and construction challenges lies in minimizing the induced ground movement, as it poses potential threats to nearby buildings and utilities along the tunnel alignment [2]. Most of the previous works were done to study the impact of the tunnel excavation on adjacent buildings which is of major interest for tunneling construction in urban areas, due to the high interaction between tunneling and existing structures [3]. Some researchers have also applied 2D numerical models for the analysis of tunnels under static conditions [4, 5]. In many of the previous studies, the impact of drilling urban tunnels on adjacent structures has been extensively studied using FE analysis and field monitoring measurements [6].

Present study aims at investigating the design parameters of an underground tunnel which is constructed in such a zone where a residential area is being developed far away from the business center, for example, near any river embankment, where new land is being developed using sand filling taken from the dredged materials from the bed of the rivers. The response of the nearby structures (bending moment, shear force) will not only be dependent on the construction sequence of the tunnel, but also will be dependent on the change in water level near the embankment. In context of the previous studies, this study aims mainly at joint response of the tunnel construction and change in water level near the embankment to other nearby structures.

2 Objectives

This study mainly deals with the behavior/response of near embankment structures (and the tunnel itself) due to tunnel construction and change in water head difference on both sides of the embankment. Specific objectives of the present research may be stated as follows:

1. To study the response of nearby road and building structures (along with the tunnel itself) due to the generated deformation.
2. To analyze the response of the tunnel and nearby structures due to change in water level on the other side of the embankment.

3. To identify suitable location for the tunnel construction by changing various variables such as depth of tunnel construction, distance of tunnel from embankment and other nearby structures.

3 Modelling Using PLAXIS-2D

In this research, a basic finite element model is developed using PLAXIS-2D. The model includes a typical soil profile, an embankment, a tunnel (constructed by a TBM), nearby structures (such as road, building etc.). Appropriate boundary and load conditions then are applied on the boundary and structures to simulate a practical situation.

A typical soil profile of Dhaka Sandy Soil (dredged fill) is used to model the soil layers (Table 1). The soil profile composed of two types of sand layer up to a depth of top 5 m, underlying a 5 m thick clay layer. Beneath the clay layer, another 10 m of sand layer is assumed. The SPT values, varying with depth are used for calculating angle of internal friction (Ø) and Modulus of Elasticity (E) with the help of empirical relations found from literature. The developed 2D plain strain model is shown in Fig. 1.

Table 1. Soil profile

Soil layer	Depth (m)	SPT (avg)	Undrained angle of internal friction, Ø' (°)	Undrained coheson, c' (kN/m^2)	Modulus of elasticity, E (kN/m^2)	Remarks
layer-1	2	11	30	0.1	15990	Cohesionless
layer-2	3	8	28	0.1	13636	Cohesionless
layer-3	5	5	25	35	21000	Cohesive
layer-4	10	11	30	0.1	15990	Cohesionless

The loading phase was defined to activate the structural loadings of two buildings having 10.0 sqm area on both sides of the center of the tunnel to be constructed and a typical road of 7.0 m width. An uniformly distributed load (UDL) of 15 kN/m^2 were applied on both of the buildings to simulate the practical situation of a typical residential building. Both the buildings and road were modelled as plate element having an average bending stiffness and normal stiffness, specified according to their thickness and concrete strength. It is to be noted here that the road were separated by a horizontal distance of 2 m from the edge of the buildings.

A total number of seven (07) calculation phases were defined for each of the trial cases. The calculation phases were defined so as to simulate the practical loading conditions sequentially. The calculation phases are stated in Table 2.

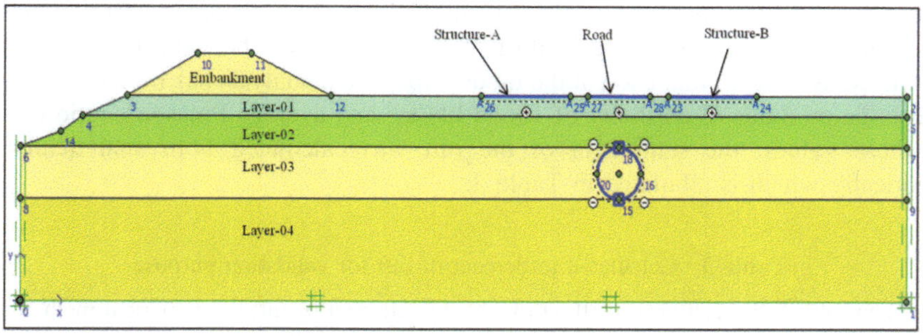

Fig. 1. Developed 2D model

Table 2. Calculation stages

Calculation phase	Name of the phase	Description of the phase
1	Mweight	Total weight of the soil is activated
2	Loading	Loads (UDL) of typical building and road are activated
3	Construction of tunnel	Construction of tunnel by TBM
4	Contraction of tunnel	2% Contraction of tunnel lining
5	WL-1	Incorporation of water head difference of 2.5 m
6	WL-2	Incorporation of water head difference of 4.5 m
7	WL-3	Incorporation of water head difference of 6.5 m

4 Validation of Model

Validation of PLAXIS-2D model of this present study has been done analytically by Poulos and Davis method (Poulos H.G. and Davis E.H., 1980).

According to Poulos and Davis, a simplified approach to calculate the settlement of a raft can be estimated by the following equation:

$$\rho = 0.947\frac{P}{B}\left[\frac{1-\nu^2}{E_s}\right]$$

Where,

ρ = settlement at the base of the raft

P = total load working on the raft

E_s = modulus of elasticity of soil below the raft

ν = Poisson's ratio of the soil below the raft

For the validation purpose of present study, a raft of 12” thickness, 10 m width has been considered with a working load of 15 kN/m^2. The soil beneath the raft has been considered as same as that used in the models for which analysis has been done, having an average modulus of elasticity of 16654 kN/m^2 and average Poisson’s ratio of 0.30. For these values, the settlement of the raft was calculated both analytically and numerically which is tabulated in Table 3.

Table 3. Calculated settlement of raft for validation purpose

Settlement of raft by analytical method of Paulos and Davis (m)	Settlement of raft by numerical method using PLAXIS-2D (m)
0.014	0.016

It is observed from Table 3 that the settlement of raft calculated by PLAXIS-2D is in well agreement with that of calculated by Poulos and Davis’s simplified method which implies that PLAXIS-2D can be used with sufficient accuracy in simulating behavior of a raft.

Another model was developed and validated comprising a shallow tunnel in a soft clay formation (Shabna P.S., Dr. N. Sankar, 2016). The major calculated stress was found approximately 346.48 kN/m^2. The maximum stress obtained at a depth of 20 m from tunnel center was validated against the obtained results which are tabulated in Table 4. In this case also, the analytical results showed to be in well agreement with the simulated results.

Table 4. Comparison of results for validation of tunnel

Constraint	Numerical calculation	Analytical calculation (PLAXIS 2D)
Maximum effective stress: $\sigma' = \sigma - u_w$	348.85 kPa	346.48 kPa

Figures 2 and 3 represents the model used and deformed mesh of the model for raft validation purpose.

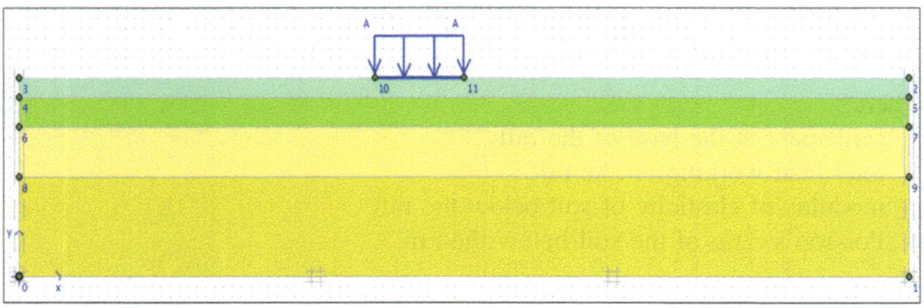

Fig. 2. Developed 2D model used for validation purpose

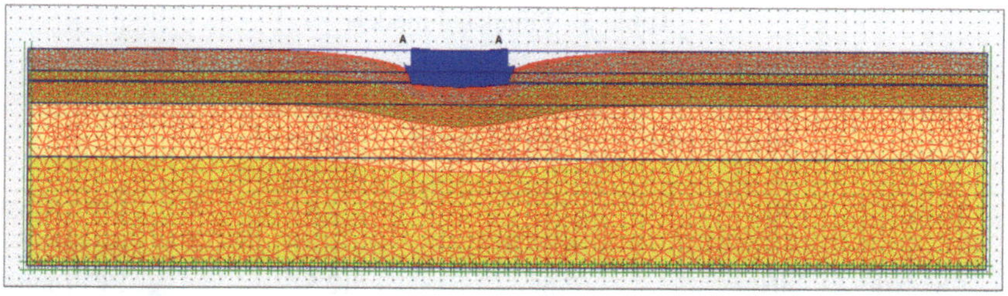

Fig. 3. Deformed mesh of the developed 2D model used for validation purpose

5 Model Results & Discussions

For this research, a total number of 09 (nine) cases were modeled and simulated, incorporating change in thickness of tunnel lining, varying horizontal position of center of tunnel, varying vertical position of center of tunnel. All these models are tabulated in Table 5.

To study the behavior of the building and road, the generated maximum displacement, moment and shear force were considered. It is note worthy to mention here that in all simulated models, two structures - A and B were similar with respect to the center of the road. For this reason only structure-B was considered for the analysis.

Results indicate that maximum displacement at the base of the road and building, both increases with increased thickness of tunnel lining (Figs. 4 and 5). Two phases are considered. One, the construction of tunnel, another is 2% contraction of tunnel lining. It is noted that during the construction of tunnel stage, the maximum deformation is less than the contraction phase because of generation of uplift forces during construction of tunnel by TBM method while in the contraction of tunnel stage, the maximum total deformation is greater because of increased overburden pressure of soil and shrinkage of tunnel lining causing larger settlement.

For better understanding, total displacement along the length of the base of Structure-B is plotted in Figs. 6 and 7 for construction of tunnel and contraction of tunnel phase respectively. From the plots it is observed that at the end of construction of tunnel phase, maximum displacement occurs near the center of structure-B for all the three cases. Displacements is less at the left corner due to generation of uplifting pressure by the TBM and at the right corner because of away from the tunnel. Whereas, at the end of contraction phase of tunnel, displacement is maximum at the left most corner due to shrinkage of tunnel lining near that point. Similarly the displacements along the base of the road are plotted in Figs. 8 and 9.

The maximum total displacement of the structure-B and road against horizontal position of the tunnel are plotted in Figs. 10 and 11 respectively. Results show that as the tunnel is shifted towards the center of structure-B, the maximum total displacement (downward) at the base of the structure decreases in the construction phase of the tunnel as during construction of the tunnel by TBM method, uplift forces are generated which tend to decrease near the center of the structure-B due to increase of downward loading (UDL). Whereas, in the contraction phase of the tunnel, the maximum total

Table 5. Simulated models

Sl. No.	Simulated case	Thickness of center of tunnel (mm)	Horizontal position of center of tunnel (m)	Vertical position of center of tunnel (m)	Remarks
01	Basic	380	67.5	12.5	15" Tunnel lining
02	Case-01	460	67.5	12.5	18" Tunnel lining
03	Case-02	510	67.5	12.5	20" Tunnel lining
04	Case-03	380	72.0	12.5	Tunnel center at +4.5 m right
05	Case-04	380	78.0	12.5	Tunnel center at +10.5 m right
06	Case-05	380	63.0	12.5	Tunnel center at −4.5 m left
07	Case-06	380	57.0	12.5	Tunnel center at −10.5 m left
08	Case-07	380	67.5	13.5	Tunnel center at +1.0 m upwards
09	Case-08	380	67.5	11.5	Tunnel center at +1.0 m downwards

displacement at the base of the structure-B increases as the tunnel is shifted towards the center of structure-B. Besides, at a zone of 5–10 m near structure-B, the rate of increase of displacement is observed to be significantly higher (Fig. 10).

The maximum total displacement at the base of the road, however, follows a different trend (Fig. 11). It tends to decrease near the center of the road in the construction phase of the tunnel due to generation of uplifting forces by TBM and there is a small zone near the center of the road near which the displacement is maximum in the contraction phase of the tunnel.

In addition to changing horizontal position of tunnel center, vertical position of it was also changed. In this article results of three cases namely Base, Case-07, Case-08 are discussed where center of the tunnel was 1 m above and below respectively for Case-07 and Case-08 than the center of tunnel in the Basic case.

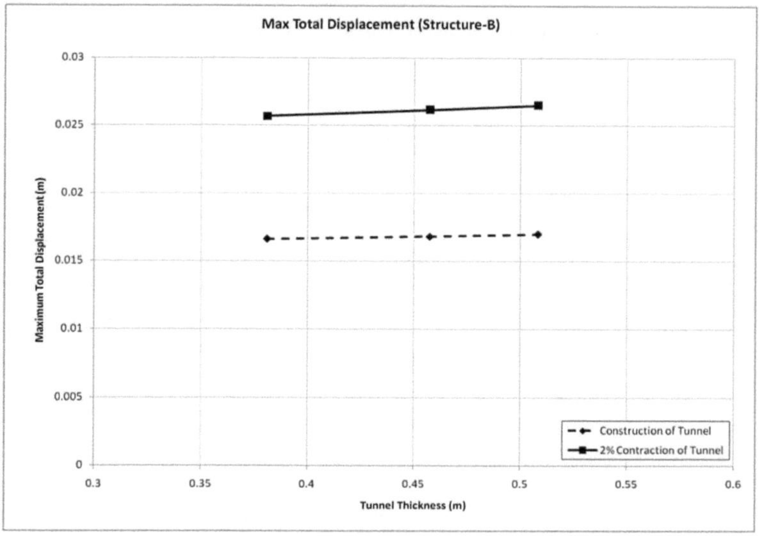

Fig. 4. Maximum total displacement of at the base of structure-B

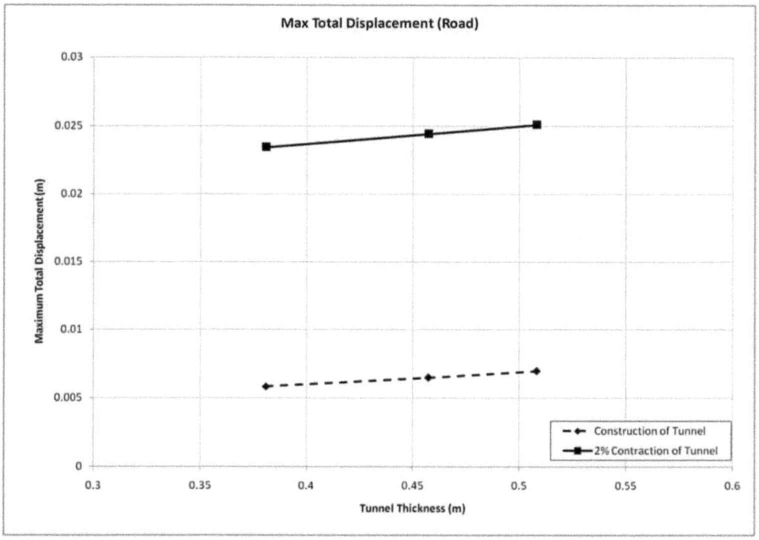

Fig. 5. Maximum total displacement of at the base of road

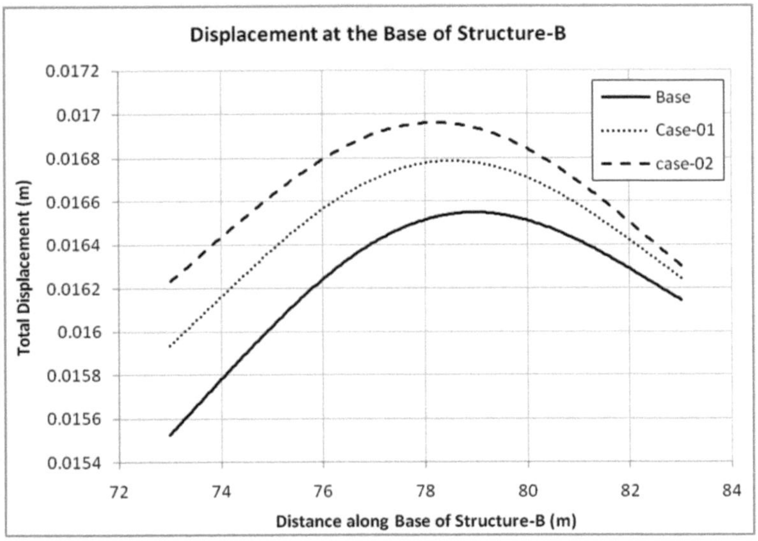

Fig. 6. Total displacement at the base of structure-B at the end of construction of tunnel

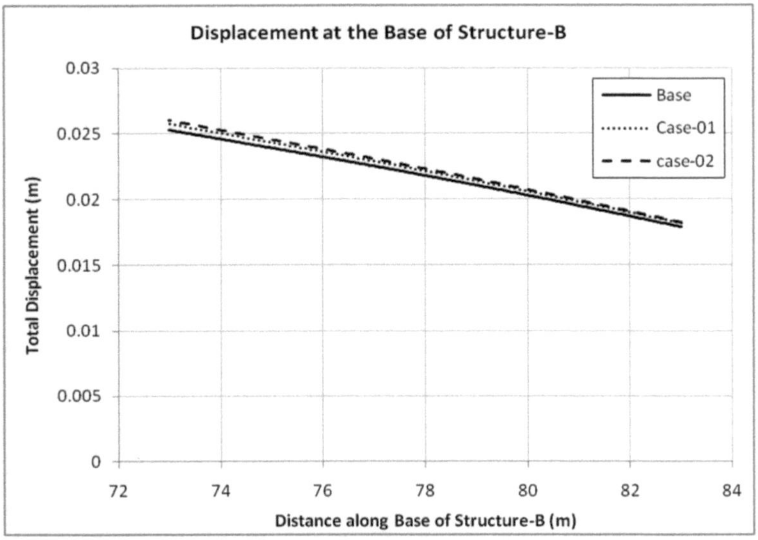

Fig. 7. Total displacement at the base of structure-B at the end of contraction of tunnel

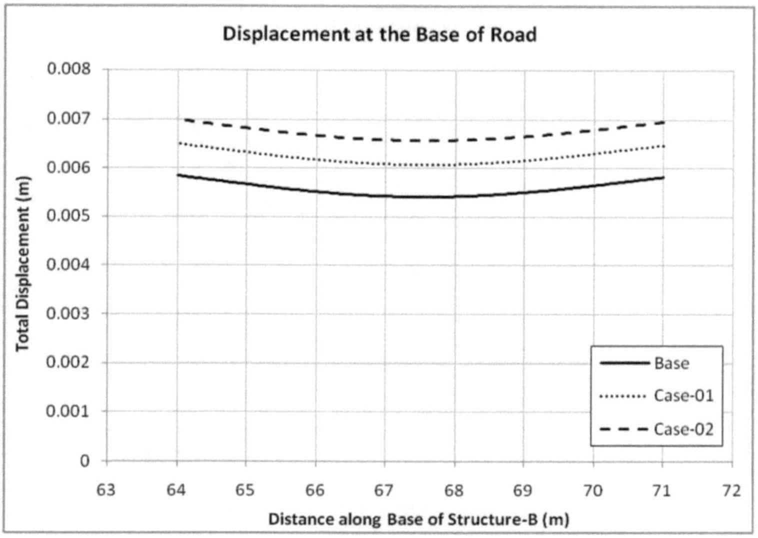

Fig. 8. Maximum displacement at the base road at the end of construction of tunnel

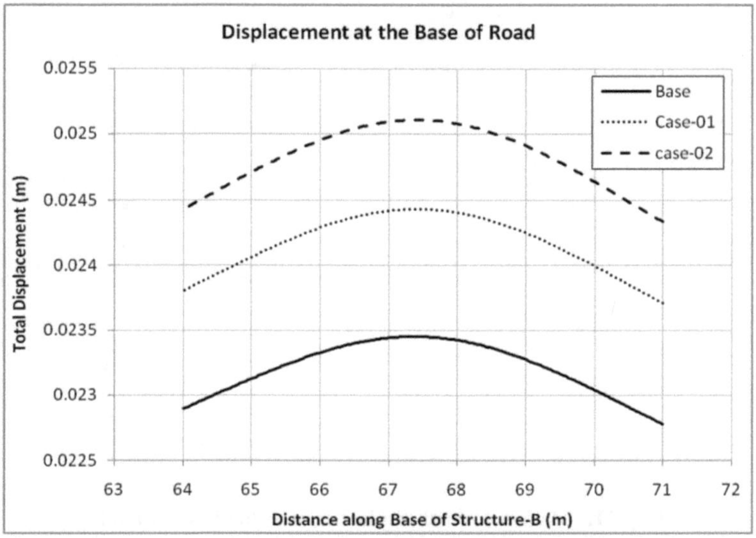

Fig. 9. Maximum displacement at the base road at the end of contraction of tunnel

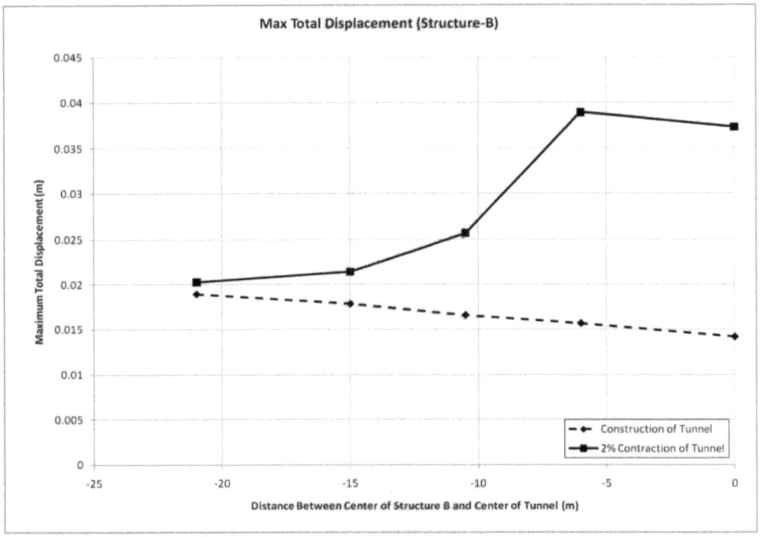

Fig. 10. Maximum total displacement at the base of structure-B

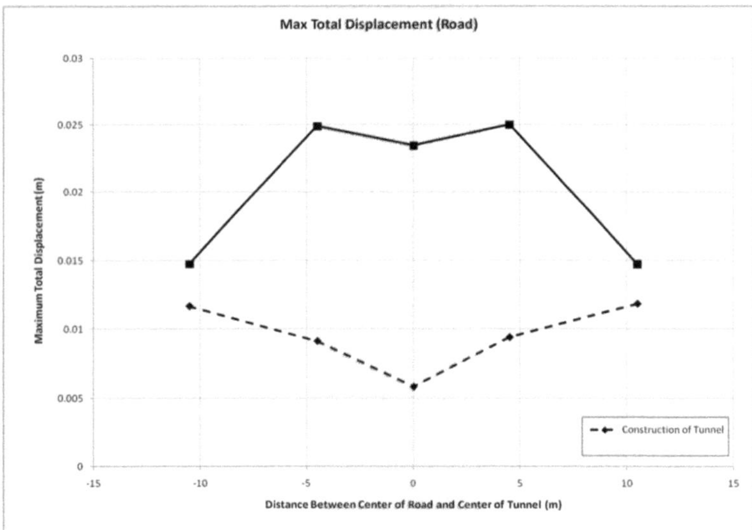

Fig. 11. Maximum total displacement at the base of road

Figure 12 shows the change in maximum displacement of structure-B for the three cases. It is observed from the figure that in the contraction phase of the tunnel, maximum displacement of structure-B increases due to increasing overburden pressure of soil when the tunnel is shifted downwards. In the construction phase the displacement

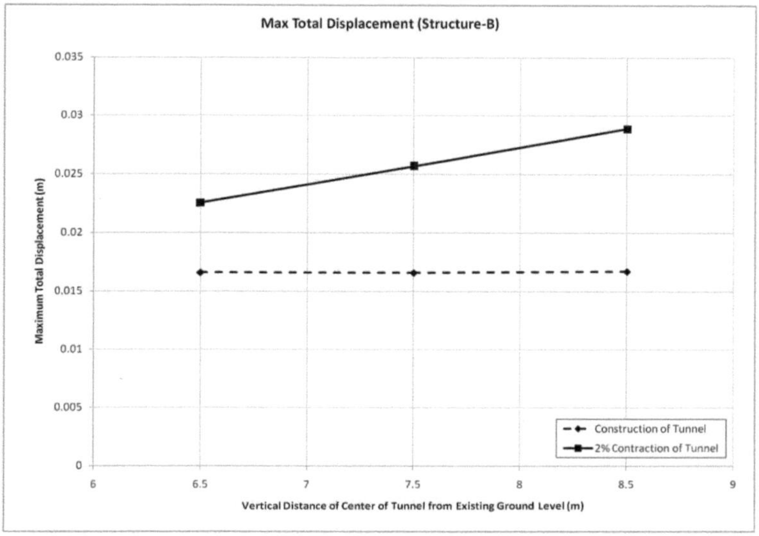

Fig. 12. Maximum total displacement of structure-B due to change in vertical position of center of tunnel

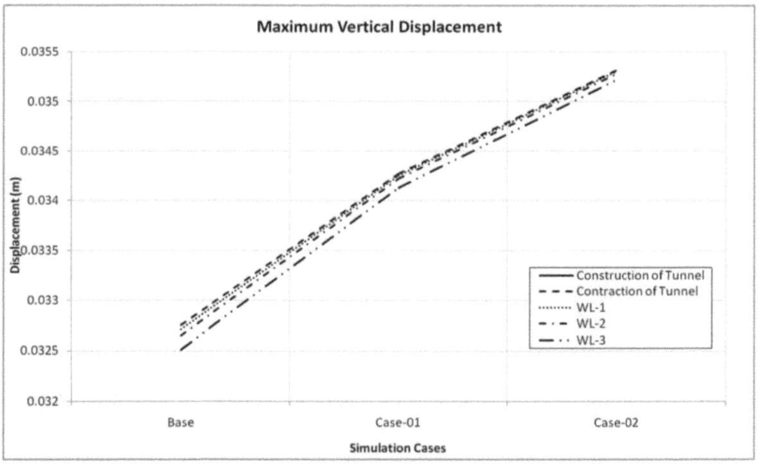

Fig. 13. Maximum vertical displacement of tunnel for change in thickness of tunnel

remains more or less constant because of generation of less amount of uplift forces by TBM as compared to increase in overburden pressure by soil downwards.

In this study, the effect of all the simulated cases on the tunnel are also analyzed. Figure 13 represent the effect of change in thickness of tunnel on maximum vertical displacement of the tunnel. It is evident from Fig. 13 that increased thickness increases displacement of the tunnel itself, mostly due to increase in self weight of the tunnel. It

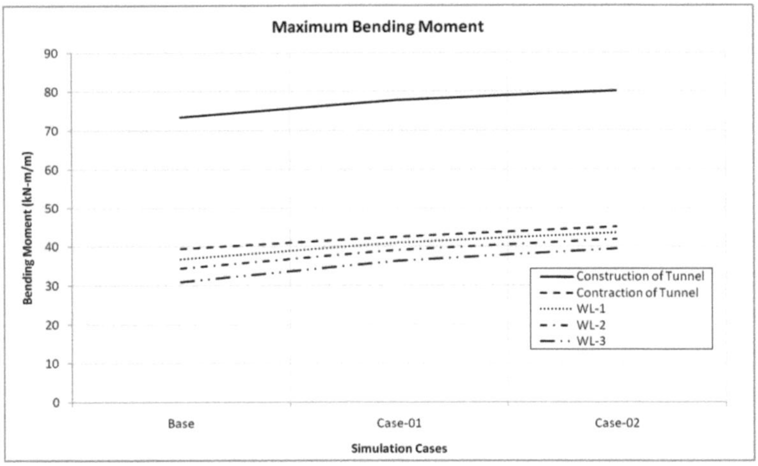

Fig. 14. Maximum bending moment of tunnel for change in thickness of tunnel

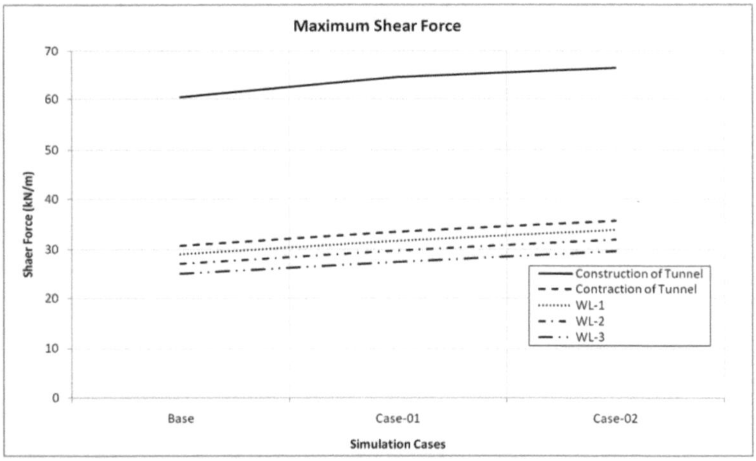

Fig. 15. Maximum bending moment of tunnel for change in thickness of tunnel

is also noted that increased water level difference on both sides of the embankment decreases the vertical displacement of tunnel. Increased displacement of the tunnel also increases the maximum bending moment and maximum shear force of the tunnel (Figs. 14 and 15).

Changing horizontal position of the tunnel also has significant effect on the displacement, bending moment and shear force of the tunnel. Figures 16, 17 and 18 represent the maximum vertical displacement, maximum bending moment and maximum shear force of the tunnel, which all shows significant increase when the tunnel is shifted horizontally towards any structure. This increase is mainly due to increase in

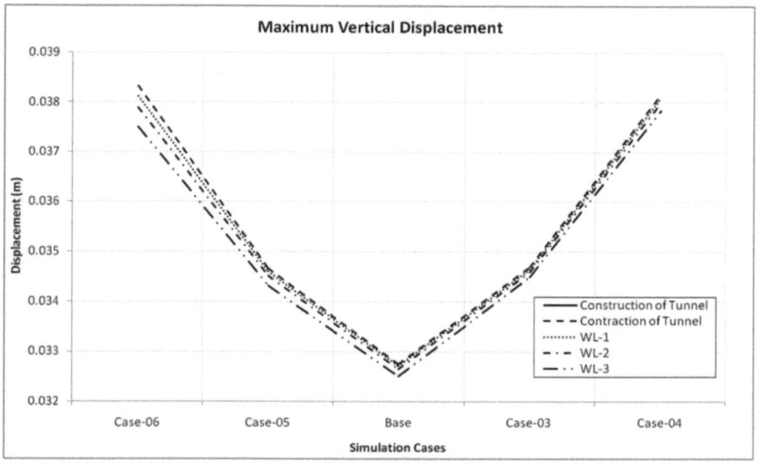

Fig. 16. Maximum vertical displacement of tunnel for change in horizontal position of tunnel

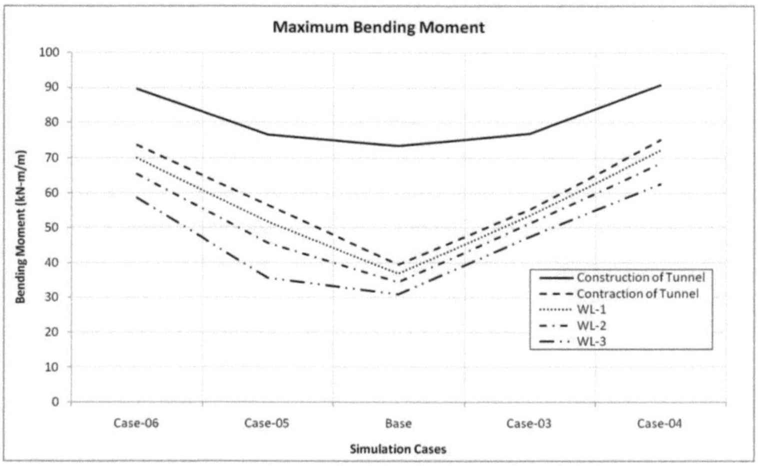

Fig. 17. Maximum bending moment of tunnel for change in horizontal position of tunnel

overburden pressure from the structures above the tunnel. In all the cases, WL-3 condition shows lowest vertical displacement of the tunnel due to generation of highest uplifting water pressure (highest water head difference).

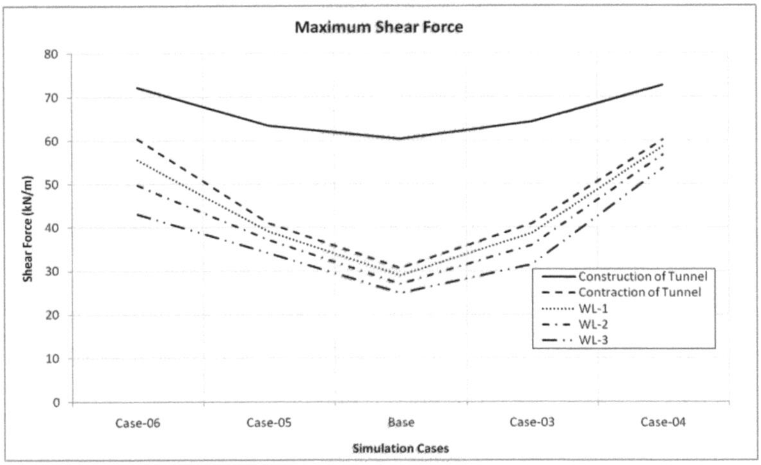

Fig. 18. Maximum shear force of tunnel for change in horizontal position of tunnel

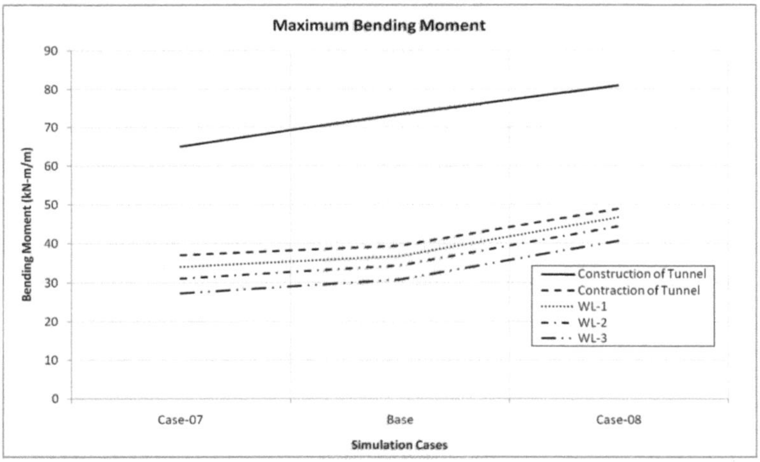

Fig. 19. Maximum bending moment of tunnel for change in vertical position of tunnel

Increased depth of construction of tunnel also shows increased stress development on the tunnel by increasing the maximum bending moment and maximum shear force of the tunnel (Figs. 19 and 20).

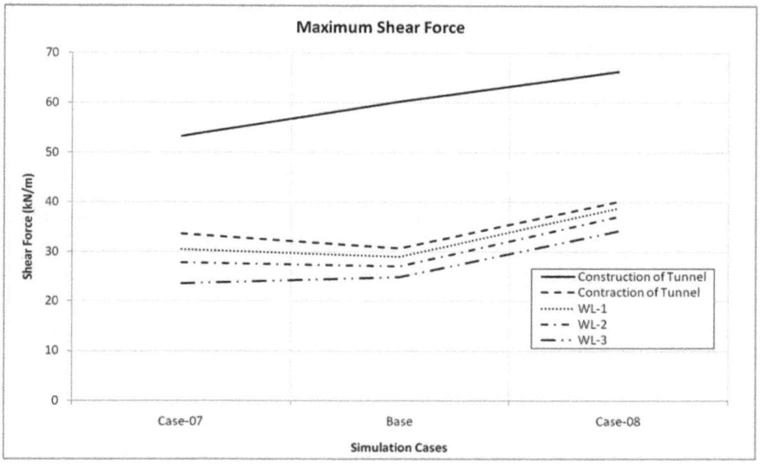

Fig. 20. Maximum shear force of tunnel for change in vertical position of tunnel

6 Conclusions

In this study, a total number of 09 (nine) 2D plain strain model were simulated to study the soil-structure interaction behavior of structures, nearby to an earthen embankment. From the obtained results and analysis done in this research, some conclusions and recommendations for future study can be made which are described in the following articles.

(a) Maximum displacement at the base of any nearby structure may increase with the increase in thickness of tunnel due to large overburden pressure above the structure.
(b) Maximum displacement of the tunnel increases significantly with increased thickness of the tunnel due to significant increase in self weight of the tunnel.
(c) The maximum vertical displacement, bending moment and shear force of the tunnel increases significantly when the tunnel is constructed very near/just below any heavily loaded structure. So, fixing horizontal position of tunnel is a matter of great importance in a crowded area.
(d) Obtained maximum bending moment and shear force of the tunnel may help the tunnel designer to a great extent in preliminary choosing best tunnel section in the most suitable horizontal/vertical position.
(e) Increased depth of tunnel causes increased stresses developed on the tunnel.
(f) Water level difference on both sides of the embankment causes seepage through the soil. This causes generation of uplifting pressures below any structure and the tunnel itself. In case of lightly loaded structures and thin tunnel sections, this may cause significant structural damage, if un-noticed.

Acknowledgements. The authors are grateful to Dr. Mehedi Ahmed Ansary, Professor, Department of Civil Engineering, Bangladesh University of Engineering and Technology (BUET), for his valuable suggestions and comments during the period of this study.

References

1. Maraš-Dragojević, S.: "Analysis of ground settlement caused by tunnel construction" (2012)
2. Sirivachiraporn, A., Phienwej, N.: Ground movements in EPB shield tunneling of Bangkok subway Project and impacts on adjacent buildings. Tunn. Undergr. Space Technol. **30**, 10–24 (2012)
3. Pickhaver, J., Burd, H., Houlsby, G.: An equivalent beam method to model masonry buildings in 3D finite element analysis. Comput. Struct. **88**, 1049–1063 (2010)
4. Desari, G.R., Rawlings, C.G., Bolton, M.D.: "Numerical modeling of NATM tunnel construction in London clay," Geotechnical Aspects of Underground Construction in Soft Ground, Rotterdam, pp. 491–496 (1996)
5. Swoboda, G.: "Finite element of new Austrian tunneling method (NATM)," In: Proceedings of the International Conference on Numerical Methods in Geomechanics, Aachen, Rotterdom, pp. 581–586 (1979)
6. Azadi, M., Zahedi, A.: "Evaluation of the impacts of drilling urban tunnels as lifelines on adjacent structures", In: Proceedings of the Twelfth East Asia-Pacific Conference on Structural Engineering and Construction (2011)
7. Poulos H.G., Davis E.H.: "Pile Foundation Analysis and Design". The University of Sydney (1980)
8. Shabna P.S., Sankar, N.: "Numerical analysis of shallow tunnels in soft ground using Plaxis2D", Int. J. Sci. Eng. Res. **7**(4) (2016)

Numerical Analysis Validation Using Embedded Pile

Esraa R. Elshehahwy[1(✉)], Ayman Eltahrany[2], and Adil Dif[1]

[1] Faculty of Engineering, Mansoura University, Mansoura, Egypt
esraa_elshehawy92@yahoo.com
[2] Faculty of Engineering, Structure Engineering Department, Mansoura, Egypt

Abstract. Piled-Raft foundation and Group piles have been widely used because of the construction of large numbers of mega projects in recent decay. This research performed the numerical 3D finite element method (FEM) analysis of piled- Raft system and Group piles in sand layers conditions of karbala soil using PLAXIS 3D. There is a need for 3D numerical modes that is able to validating a real field test. Verification problem is simulated from real field test of four piled-Raft system by using PLAXIS 3D program. The soils are modelled with small hardening soil model while the raft was modelled by using plate and embedded pile modelled the piles. The results of numerical analysis shows that there is an excellent matching between field test and the FEM, also, shows that the ratio between the summation of load carried by all piles and the total load on the raft foundation which is defined by load sharing ratio of piles.

Keywords: Piles raft · Pile group · Validation · Finite element method Embedded pile · PLAXIS3D

1 Introduction

Due to the rapid development in using tall buildings with excessive loads, multi-numbers of mega projects is using deep foundation as an economic foundation type to be applied in different geotechnical conditions. (O'Neil 2006) indicate that the 3D dimensional in nature and the complex behavior in the interaction between its components are the reason of the complex behavior of piled raft simulation in the numerical analysis. In the present paper will focus on the load- settlement curve analysis of piled foundation using finite element based on the real field test. Standared penetration test is one of the most important test in situ used in cohessionless soil like sand layers. Hussieen et al. (2017) used it to determine the soil profile based on its results from site investigation.

Even if the load applied on the head of the pile is with low values, the response of load-settlement curve will be nonlinear (Vesic 1970; O'Neil and Reese 1972; O'Neil and Hawkins 1980). SO, we have to propose some assumption to overcome the nonlinearity in pile group analysis. (Poulos 1989; Clemente 1990; Lee and Xiao 2001; Castelli and Meugeri 2002; Randolph 2003; Klar and Leung 2009). Due to the rapid development in technologies of computer, it is became easy to solve complex problems and saving in time and computational effort.

H. Shehata and C. S. Desai (Eds.): GeoMEast 2018, SUCI, pp. 199–208, 2019.
https://doi.org/10.1007/978-3-030-01926-6_15

2 Piled Raft Foundation and Pile Group Concept

The concept of piles construction is used to reduce the average settelement and the differential settlements. Burland et al. (1977). Therefore, Piled raft foundation was chosen because of:

- The great capacity of structure on soil can carried
- the soil capacity is undergo great stresses.

The piled raft foundation consists of three main components: soil, raft and pile. These components related to each other by soil- structure interactions like: soil-pile interaction, soil-raft interaction and pile-raft interaction as presented in Fig. 1 (Poulos 2011). Otherwise the pile group, its interactions are between soil-pile and pile-raft only and that is the main reason differs between Piled-Raft system and Pile Group.

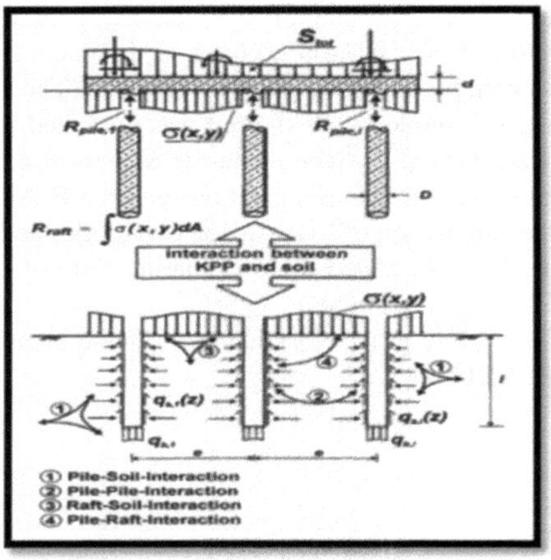

Fig. 1. Poulos 2011

Within applying a vertical load to piled-raft foundation,, the load is transferred to the ground by both tracks, the pile and the raft in piled-raft foundation system. So, the total vertical load is sharing between piles and raft. Otherwise pile group, the load is transmitted to the ground by piles.

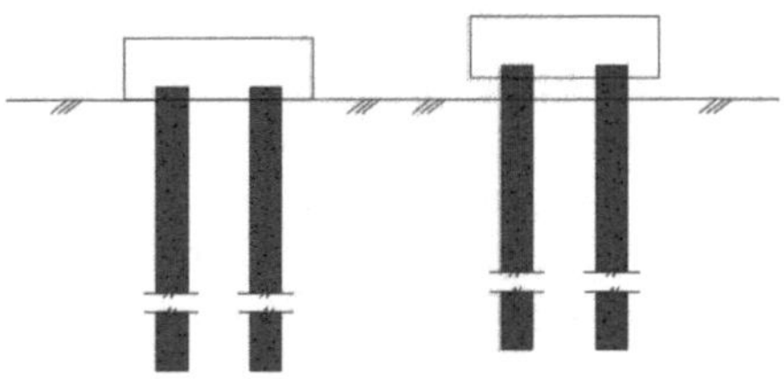

Fig. 2. The difference between piled raft and pile group

Generally, this is the main difference between pile group and piled-raft foundation. As shown in Fig. 2 there is a clearance between raft and ground surface in pile group system.

Consequently, the piled-raft foundation has some features over the pile group to be economical method and has a good serviceably as following:

- at the same design requirements, piled-raft foundation requires a fewer piles than pile group
- Poulos et al. (2011) presented at their studies that the raft redistributes the load to the remained piles, if there is a pile will be damaged or has a defect.
- Clancy and Randolph, 1993 reached to the percentage which the raft carried to transmit to subsoil which is 30%: 50%

the stress applied from raft to subsoil increases the lateral stress between soil and piles and that increases the bearing capacity of piled-raft foundation than pile group (Katzenbach et al. 1998).

3 Soil Investigation

In this study, the soil profile is located on Karbala in Iraq – Karbala city, 100 km south west of Baghdad which the area was worked on by Hussein et al. (2017). From engineering geological layers investigation, two geological layers were generalized as follows:

1. Very dense sand (the depth is up to 5 m from ground surface).
2. medium dense sand (the depth is up to 15 m from ground surface)

These two layers were used in the verification of the real field test. The ground water table is at 15 m level from underground surface.
From site investigation, soil properties were determined as in Fig. 3. geological investigation involved the number of bowels in Standard Penetration Test (SPT), water content for every layer and two boreholes as shown in Fig. 4.

A static load field test, according to ASTM D1143 (2007), was applied on Karbala's soil in Iraq to get a good result. Load was compressed by using jacks, the capacity of one jack is 500 ton applied on a raft with (1.8 m × 1.8 m × 0.6 m) dimensions. Consequently, The Fig. 5 show the load- settlement curve from field test. Hussein et al. (2017).

4 3d Finite Element Modelling

Piled-raft foundation system is a 3D complex problem, which needs 3D finite element method to be solve. Consequently, we use PLAXIS 3D 2013 to simulate the structure numerical models. Numerical model geometry consists of three material components: (1) soil elements; (2) embedded pile elements; and (3) plate element for pile cap.

A quarter of model was used to reduce the computational time and effort because, PLAXIS 3D has a symmetrical around x-axis and around y-axis. So, reducing the

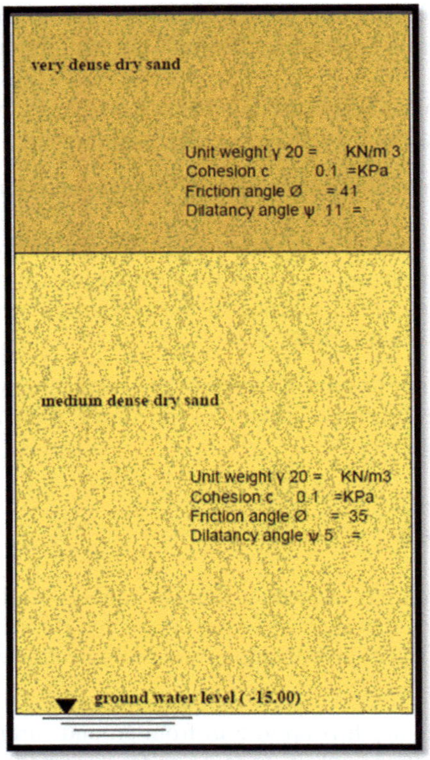

Fig. 3. Soil profile

elements numbers might save more time. Figure 5 shows a typical three dimensional finite element mesh of 2 × 2 piles used in this study in the both piled raft and pile group (Fig. 6).

Embedded pile, as in Fig. 7, was selected to model the four (2 × 2) piles in which the piles are assumed to be like slender beam element. Plate element was selected to model pile cap because the plate element and embedded pile have six degree of freedom and that allows completely compatible connection with each other. There are many constitutive models simulated the soil behavior, the hardening soil model with small-strain stiffness (HSsmall) was chosen to simulate the non-linear stress-strain of the cohesionless soil.

SOIL MODEL AND PARAMETER.

The constitutive model for two layers of sand is hardening soil model material in drained condition. HSsmall hardening model requires more than thirteen parameters. the Table 1 presented the HS small model' s parameters with their standard units.

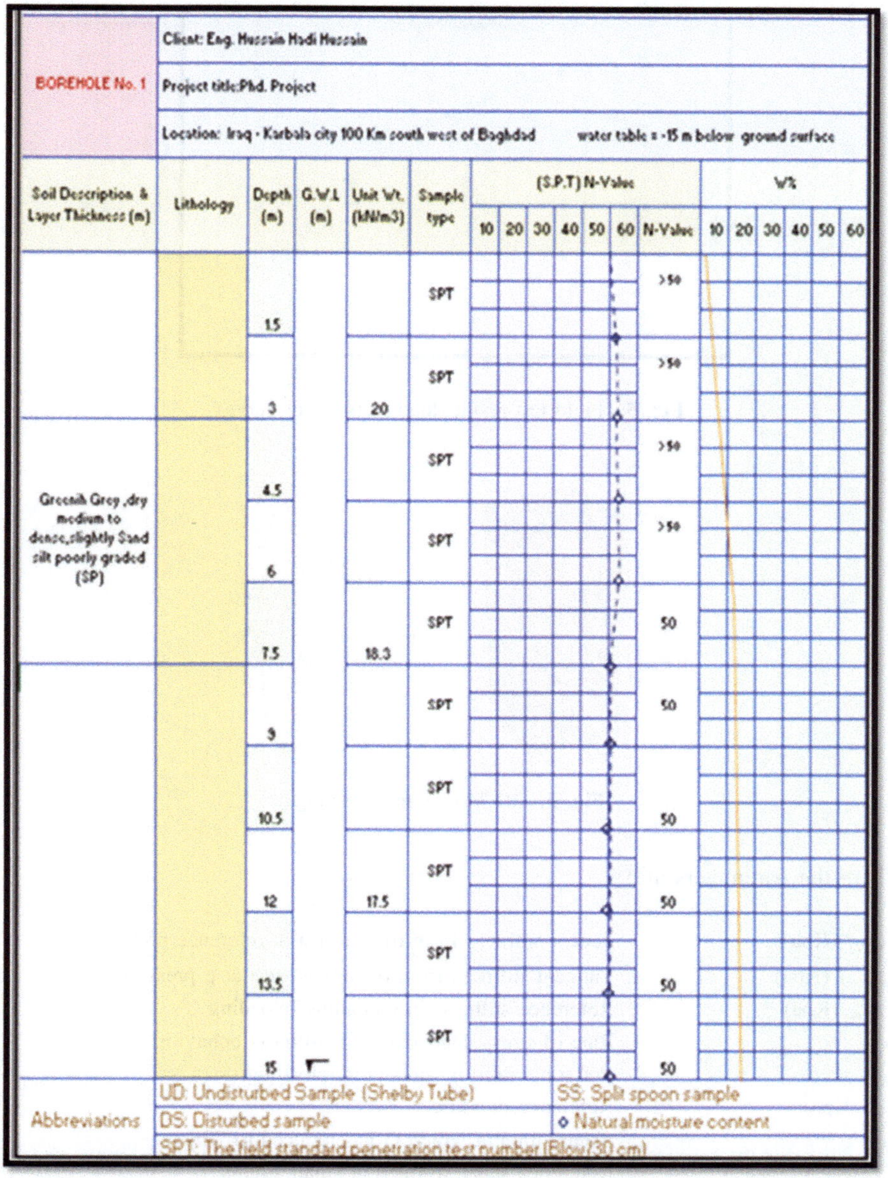

Fig. 4. Description of the boreholes. Hussein et al. (2017)

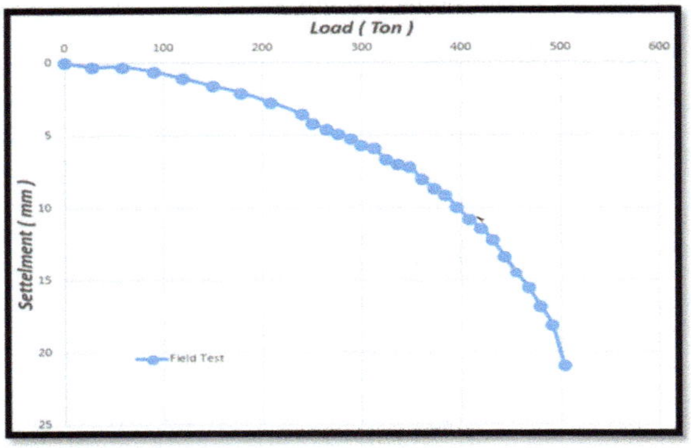

Fig. 5. Field test result (load- settlement curve)

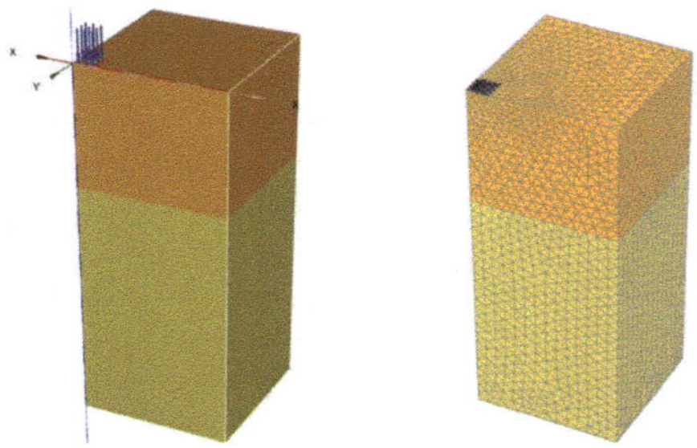

Fig. 6. 3D Model and meshing

Where the paremeters are:

$E_{50.ref}$ (Kpa)	Secant stiffness from triaxial test at reference pressure
$E_{oed.ref}$ (Kpa)	Tangent stiffness from oedometer test at p pressure
$E_{ur.ref}$ (Kpa)	Reference stiffness in unloading/reloading
m	Rate of stress dependency in stiffness behavior
ν_{ur}	Poisson's ratio in unloading/reloading
P_{ref} (Kpa)	Reference pressure
$\gamma_{0.7}$	Shear strain at which G has reduced to 70%
$G_{0.ref}$	Reference shear stiffness at small strains
c (Kpa)	Cohesion
Ø	Friction angle
ψ	Dilatancy angle
$K0_{NC} = 1 - \sin\emptyset$	Stress ratio $\sigma\sigma'xx/\sigma'yy$ in 1 D primary compression

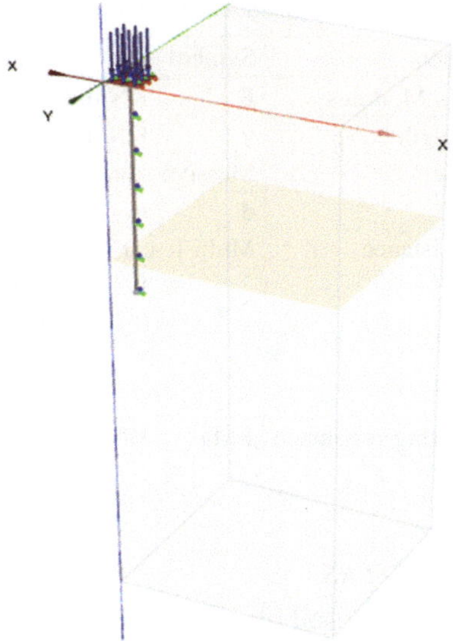

Fig. 7. Embedded pile

Table 1. Material properties of sand soil model (HS small)

Property	Layer 1: very dense sand	Layer 2: medium dense sand
Unit weight γ (KN/m3)	20	20
Drainge type	Drained	Drained
$E_{50.ref}$ (Kpa)	60000	35000
$E_{oed.ref}$ (Kpa)	60000	35000
$E_{ur.ref}$ (Kpa)	180000	105000
m	0.4	0.5
ν_{ur}	0.15	0.15
P_{ref} (Kpa)	100	100
$\gamma_{0.7}$	0.15E-4	0.15E-4
$G_{0.ref}$	130000	100000
Cohesion c (Kpa)	0.1	0.1
Friction angle Ø	41	35
Dilatancy angle ψ	11	5
Tension cut-off (Kpa)	0	0
$K0_{NC} = 1 - \sin \emptyset$	0.344	0.426
$K0_{ini} = K0_{NC}$	0.344	0.426

Table 2. Embedded pile parameters.

Parameter	Symbol	Unit	Value
Young's Modulus	**E**	KN/m2	2E+07
Unit weigth	γ	KN/m3	25
Pile type	Massive square pile		
Width	**d**	m	0.3
Skin resistance	Multi-linear		
		L (m)	T (KN/m)
		0	0
		5	144
		6	120
End-bearing resistance	**FMax**	KN	1296

EMBEDDED PILE PARAMETERS.

The embedded pile intersects with soil by special interface element that include the pile skin friction and pile end-bearing resistance. The embedded pile parameters is summarized in Table 2. Also, the length of the pile is 6 m with spacing is equal to three time its width. The skin friction resistance and end-bearing resistance are based on static manual calculation of pile load capacity.

5 Computed Results

PILE LOAD-SETTELEMENT CURVE OF THE PILE FOUNDATION.

Figure 8 illustrated the computed load settlement curve for piled raft foundation simulated the real piled-raft field test. The coefficient of horizontal soil stress K for

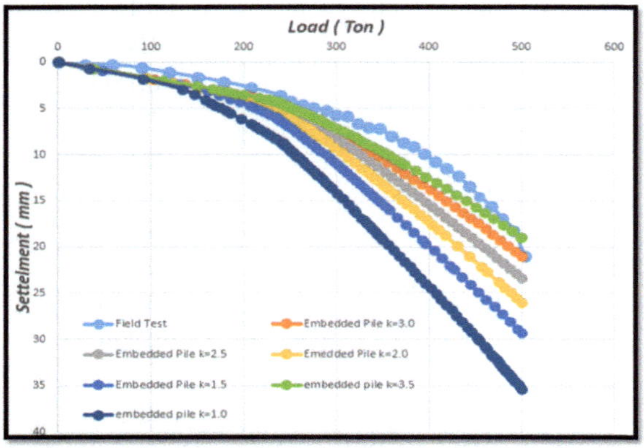

Fig. 8. Load- settlement curve with different K

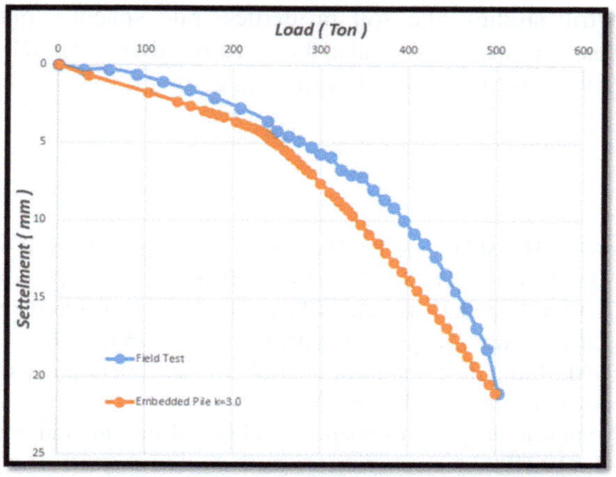

Fig. 9. Load- settlement curve

driven piles differs from bored piles and range of K equal from 1 to 2 or more in cohesiveless soil, Tomlinson (2015).

Consequently, that is what was shown. So, the best coefficient of horizontal soil stress matching with the real test field is K = 3.0, Fig. 9 represents that. The maximum settlement in simulation the piled raft foundation is 21.01 mm and that has an excellent matching with the field test which the settlement reached to 21 mm in field. So, the agreement of the results are achieved.

The load was applied on the raft is equal 500 ton, the pile in numerical analysis can carry normal force equal to 673 KN, which was around 53.84% from total load. The remaining 46.16% from total load is transferred directly from the raft to the soil by contact. and the 46.16% is acceptable percentage as Clancy and Randolph shown.

6 Conclusions

This paper presents numerical analysis of 3D finite element modeling of piled raft foundation subjected to vertical load. Numerical studies employed using embedded pile element by 3D finite element analysis. The accuracy of embedded pile element is validated by comparing the results of the 3D finite element analysis and real field test and a number of conclusion can be presented:

- Increasing K in the cohesionless soil to high value, introduced and excellent match with the real field test.
- PLAXIS 3D is considered very good program in simulation the foundation depending on its type.
- The raft can transfer around 46.16%, in the case presented in paper, to the subsoil by contact

Moreover, parametric studies like soil properties, pile spacing, pile diameter, pile length and thickness of raft, should also be performed by PLAXIS 3D to see the behavior of pile load distribution in pile-raft foundation.

References

Hussein, H.H., Karim, H.H., Shlash, K.T.: Theoretical verification for full-scale tests of piled Raft Foundation. In: Abu-Farsakh, M., Alshibli, K., Puppala, A. (eds.), Advances in Analysis and Design of Deep Foundations, GeoMEast 2017. Sustainable Civil Infrastructures. Springer, Cham (2017). https://doi.org/10.1007/978-3-319-61642-1_16

Casteli, F., Maugeri, M.: Simplified nonlinear analysis for settlement prediction of pile groups. J. Geotech. Geoenviron, Eng. **128**(1), 76–84 (2002)

Clemente, J.L.M.: Nonlinear analysis of settlement and load distribution in pile groups. Comput. Struct. **36**(2), 203–210 (1990)

Klar, A., Leung, Y.F.: Simple energy- based method for nonlinear analysis of incompressible pile groups inn clays. J. Geotech. Geoenviron. Eng. **135**(7), 960–965 (2009)

Lee, K.M., Xiao, Z.R.: A simplified nonlinear approach for pile group settlement analysis in multilayered soils. Can. Geotech. J. **38**(5), 1063–1080 (2001)

Poulos, H.G.: Pile behavior-theory and application. Geotechnique **39**(3), 365–415 (1989)

Randolph, M.F.: RATZ manual version 4.2:Load transfer analysis of axially loaded piles. University of Western Australia (2003)

O'Neil, M.W., Reese, L.C.: Behavior of bored piles in Beaumont clay. J. Soil Mech. Found. Div. **98**, 195–213 (1972)

O'Neil, M.W., Hawkins, R.A.: Field study of pile group action- summary of results. In: Proceedings of Pile Group Prediction Symposium, University of Mayrland, Collage Park, MD (1980)

Vesic, A.S.: Tests on instrumented piles, Ogeechee River site. J. Soil Mech. Found. Div. ASCE **96**, 561–584 (1970)

O'Neil, M.W.: Piled rafts in geotechnical practice—an international perspective. In: Proceedings of 10 th International Conference on Piling and Deep Foundations, Amsterdam, pp. 197–208 (2006)

Burland, J.B., Brooms, B.B., De Mello, V.F.B.: Behavior of foundation and structures. In: Proceedings of the 9 th International Conference on Soil Mechanics and Foundation Engineering, Tokyo, Japan, vol. 2, pp. 495–546 (1977)

Poulos, H.G., Small, J.C., Chow, H.: Piled raft foundations for tall buildings. Geotech. Eng. J. SEAGS AGSSEA **42**(2), 78–84 (2011)

Katzenbach, R., Arslan, U., Moorman, C., Reul, O.: Piled Raft Foundation: interaction between piles and raft. Darmst. Univ. Technol. **4**, 279–296 (1998)

Tomlinson, M.J: Foundation Design and Construction, 7th edn. Pitman, London (2015); Swiss Standard SN 670 010b: Characteristic Coefficients of Soils. Association of Swiss Road and Traffic Engineers, Zurich (1999)

PLAXIS:Vermeer and Brinkgreve, Manual (2013)

Author Index

H. Shehata and C. S. Desai (Eds.): GeoMEast 2018, SUCI, p. 209, 2019.
https://doi.org/10.1007/978-3-030-01926-6

MIX
Papier aus verantwortungsvollen Quellen
Paper from responsible sources
FSC® C105338

If you have any concerns about our products,
you can contact us on
ProductSafety@springernature.com

In case Publisher is established outside the EU,
the EU authorized representative is:
Springer Nature Customer Service Center GmbH
Europaplatz 3, 69115 Heidelberg, Germany

Printed by Libri Plureos GmbH
in Hamburg, Germany